Lecture Notes in Computer Science   11168

Commenced Publication in 1973
Founding and Former Series Editors:
Gerhard Goos, Juris Hartmanis, and Jan van Leeuwen

Editorial Board

More information about this series at http://www.springer.com/series/7407

Shichao Zhang · Tie-Yan Liu
Xianxian Li · Jiafeng Guo
Chenliang Li (Eds.)

Information Retrieval

24th China Conference, CCIR 2018
Guilin, China, September 27–29, 2018
Proceedings

Editors
Shichao Zhang
School of Computer Science
  and Information Technology
Guangxi Normal University
Guilin
China

Tie-Yan Liu
Microsoft Research Asia
Beijing
China

Xianxian Li
School of Computer Science
Guangxi Normal University
Guilin
China

Jiafeng Guo
Chinese Academy of Sciences
Beijing
China

Chenliang Li
Wuhan University
Wuhan
China

ISSN 0302-9743   ISSN 1611-3349 (electronic)
Lecture Notes in Computer Science
ISBN 978-3-030-01011-9   ISBN 978-3-030-01012-6 (eBook)
https://doi.org/10.1007/978-3-030-01012-6

Library of Congress Control Number: 2018955274

LNCS Sublibrary: SL1 – Theoretical Computer Science and General Issues

This Springer imprint is published by the registered company Springer Nature Switzerland AG
The registered company address is: Gewerbestrasse 11, 6330 Cham, Switzerland

Preface

The 2018 China Conference on Information Retrieval (CCIR 2018), co-organized by the Chinese Computer Federation (CCF) and the Chinese Information Processing Society of China (CIPS), was the 24th installment of the conference series. The conference was hosted by Guangxi Normal University in Guilin, Guangxi, China, during September 27–29, 2018.

The annual CCIR conference serves as the major forum for researchers and practitioners from both China and other Asian countries/regions to share their ideas, present new research results, and demonstrate new systems and techniques in the broad field of information retrieval (IR). Since CCIR 2017, the conference has enjoyed contributions spanning the theory and application of IR, both in English and Chinese.

This year we received 161 submissions from both China and other Asian countries, among which 53 were English papers and 108 were Chinese ones. Each submission was carefully reviewed by at least three domain experts, and the Program Committee (PC) chairs made the final decision. The final English program of CCIR 2018 featured 22 papers.

CCIR 2018 included abundant academic activities. Besides keynote speeches delivered by world-renowned scientists from China and abroad, as well as traditional paper presentation sessions and poster sessions, we also hosted doctoral mentoring forums, a young scientist forum, an evaluation workshop, and tutorials on frontier research topics. We also invited authors from related international conferences (such as SIGIR, WWW, WSDM, CIKM) to share their research results as well.

CCIR 2018 featured four keynote speeches: "Exploring a Cyber/Physical/Social Model of Context" by Mark Sanderson (RMIT University), "Toward Building Self-Training Search Engines" by Hang Li (Toutiao), "Natural Language Processing R&D in Alibaba" by Luo Si (Alibaba), and "Dynamic Search and Beyond" by Grace Hui Yang (Georgetown University).

The conference and program chairs of CCIR 2018 extend their sincere gratitude to all authors and contributors to this year's conference. We are also grateful to the PC members for their reviewing effort, which guaranteed that CCIR 2018 could feature a quality program of original and innovative research in IR. Special thanks go to our sponsors for their generosity: Microsoft, Huawei, Chancein, Baidu, Bytedance, and Sogou. We also thank Springer for supporting the best paper award of CCIR 2018.

August 2018

Shichao Zhang
Tie-yan Liu
Xianxian Li
Jiafeng Guo
Chenliang Li

Organization

Steering Committee

Shuo Bai	Shanghai Stock Exchange, China
Xueqi Cheng	Institute of Computing Technology, Chinese Academy of Sciences, China
Shoubin Dong	East China University of Science and Technology, China
Xiaoming Li	Beijing University, China
Hongfei Lin	Dalian University of Technology, China
Ting Liu	Harbin Institute of Technology, China
Jun Ma	Shandong University, China
Shaoping Ma	Tsinghua University, China
Shuicai Shi	Beijing TRS Information Technology Co., Ltd., China
Mingwen Wang	Jiangxi Normal University, China

Conference General Chairs

Shichao Zhang	Guangxi Normal University, China
Tieyan Liu	Microsoft Research Asia, China

Program Committee Chairs

Xianxian Li	Guangxi Normal University, China
Jiafeng Guo	Institute of Computing Technology, Chinese Academy of Science, China

Youth Forum Chairs

Zhicheng Dou	Renmin University of China, China
Zhumin Chen	Shandong University, China

Industry Forum Chairs

Zhi Li	Guangxi Normal University, China
Ming Chen	Guangxi Normal University, China

Publication Chair

Chenliang Li	Wuhan University, China

Publicity Chair

Chenliang Li   Wuhan University, China

Organizing Chairs

Zhenjun Tang   Guangxi Normal University, China
Zhixin Li   Guangxi Normal University, China
Jingli Wu   Guangxi Normal University, China

Organizing Committee

Jinlu Liu   Guangxi Normal University, China
Yonggang Li   Guangxi Normal University, China
Jinyan Wang   Guangxi Normal University, China
Peng Liu   Guangxi Normal University, China
Lie Wang   Guangxi Normal University, China
Yangding Li   Guangxi Normal University, China
Shenglian Lu   Guangxi Normal University, China

Program Committee

Shuo Bai   Shanghai Stock Exchange, China
Xueqi Cheng   Institute of Computing Technology, Chinese Academy
                                    of Sciences, China
Fei Cai   National University of Defense Technology, China
Dongfeng Cai   Shenyang Aerospace University, China
Zhumin Chen   Shandong University, China
Yi Chang   Huawei Technologies Co., Ltd., China
Yajun Du   Xihua University, China
Shoubin Dong   South China University of Technology, China
Zhicheng Dou   Renmin University of China, China
Shicong Feng   Beijing Siming Software System Co., Ltd., China
Jiafeng Guo   Institute of Computing Technology, Chinese Academy
                                    of Sciences, China
Xuanxuan Huang   Fudan University, China
Yu Hong   Suzhou University, China
Minlie Huang   Tsinghua University, China
Zhongyuan Han   Heilongjiang Institute of Technology, China
Donghong Ji   Wuhan University, China
Tieyan Liu   Microsoft Asia Research Institute, China
Yiqun Liu   Tsinghua University, China
Ting Liu   Harbin Institute of Technology, China
Hang Li   Toutiao AI Lab, China
Hongfei Lin   Dalian University of Technology, China

Contents

Information Extraction and Sentiment Analysis

Social Computing

Recommendation

Sequence Modeling

Query Processing and Retrieval

Translating Embeddings for Modeling Query Reformulation

Rongjie Cai, Yiqun Liu[✉], Min Zhang, and Shaoping Ma

Department of Computer Science and Technology,
Beijing National Research Center for Information Science and Technology,
Tsinghua University, Beijing 100084, China
jayjay7@163.com, {yiqunliu,z-m,msp}@tsinghua.edu.cn
http://www.thuir.cn

Abstract. Query reformulation understanding is important for Information Retrieval (IR) tasks, such as search results reranking and query recommendation. Conventional works rely on the textual content of queries to understand reformulation behaviors, which suffer from data sparsity problems. To address this issue, We propose a novel method to efficiently represent the behaviors of query reformulation by the translating embedding from the original query to its reformulated query. We utilize two-stage training algorithm to make the learning of multilevel intentions representation more adequate. We construct a new corpus of shopping search query log and create a query reformulation graph based on this dataset. Referring to knowledge graph embedding methods, we use the accuracy of intentions prediction to evaluate experimental results. Our final result, an increase of 20.6% of the average prediction accuracy in 21 intentions, shows significant improvement compared to baselines.

Keywords: Translating embeddings · Query reformulation
Knowledge graph

1 Introduction

While performing complex search tasks, search users usually need to submit more than one query to search systems. Understanding how users reformulate queries is as important as the users' examination behavior study [1] for IR related researches. Existing works show that query reformulation behaviors contain important signals of users satisfaction perception with search results [8] and their actual information needs [9]. Uncovering the semantic relationship behind reformulated queries may also help search engine to better rank results. Prior works have made much progress in interpreting query reformulation actions. For example, Huang et al. create a taxonomy of query refinement strategies and

---

R. Cai—This work is supported by Natural Science Foundation of China (Grant No. 61622208, 61732008, 61532011) and National Key Basic Research Program (2015CB358700).

S. Zhang et al. (Eds.): CCIR 2018, LNCS 11168, pp. 3–15, 2018.
https://doi.org/10.1007/978-3-030-01012-6_1

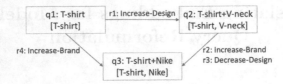

Fig. 1. Example of a query reformulation graph.

build a rule-based classifier to detect each type of reformulation [2]. Sloan et al. present a term-based methodology and provide valuable insight into understanding query reformulation [3]. However, most of these efforts rely on the textual content of queries. The average length of queries is usually no more than a few words, which makes this kind of efforts suffer from data sparsity problems.

Recently, many works attempt to embed a knowledge graph into a continuous vector space, called knowledge graph embedding, to aggregate global knowledge over a graph. For example, TransE [10] interprets relations as translation operating on the low-dimensional embeddings from head entities to tail entities. TransH [11] and TransR [12] improve embedding algorithm by modeling relations as translation on a hyperplane and a relation space respectively. Inspired by these recent efforts, we add translation module to our method, which will help the model to better understand query reformulation intentions.

We use an example to interpret how translation embedding methods work on query reformulation intentions understanding. There are two users' query logs as follows.

- User1: T-shirt, T-shirt+V-neck, T-shirt+Nike
- User2: T-shirt, T-shirt+Nike

According to the query reformulation by users, we construct query graph shown in Fig. 1, which treat queries as nodes and reformulation intentions as edges. The structure of query reformulation graph is similar to conventional knowledge graph, where queries are entities and intentions are relations. Adopting knowledge graph embedding methods to analyze query reformulation, we can predict the reformulation intentions by calculating the difference of next query vectors and current query vectors.

However, the accuracy of traditional knowledge graph embedding methods is not very good for intention prediction, which can't be applied directly to query reformulation understanding. Thus, we redefine this problem by combining translation module and classification module in our method. In the vertical search, users' query reformulation intentions can't be described simply by one-level relationship. To meet the characteristics of query data, we define 3 first-level relations and 10 second-level relations to represent intentions. We collect query log from shopping search engine and create query reformulation graph by treating queries as entities and intentions as relations. A novel two-stage training method is proposed in our work, and experimental results show significant

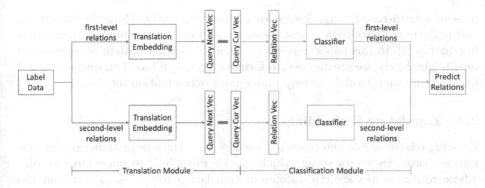

Fig. 2. The framework of our model.

improvement compared to state-of-the-art baselines including TransE, TransH and TransR.

To summarize, we make the following contributions:

- Inspired by knowledge graph embedding methods, we define query reformulation understanding as a problem of intentions embedding.
- A new framework is proposed to solve the problem of multilevel intentions, which frequently exist in query reformulation logs.
- We construct a new corpus of shopping search query reformulation and verify the effectiveness of our method based on this dataset.

2   Related Work

2.1   Query Reformulation

Query reformulation is the process that queries issued to search engines are modified after examination of the results, which is an important topic in web search. Understanding users' reformulation behaviors accurately will help search engine to improve performance. Jiang et al. propose a supervised approach to query auto-completion based on the analysis of users' reformulation behaviors [4]. Santos et al. create a probabilistic framework for Web search result diversification and exploit query reformulation as a means to uncover different query aspects [7]. Hassan et al. study additional implicit signals based on the relationship of adjacent queries and predict the satisfaction of non-clicking users [8]. Prior works show that the analysis of query reformulation plays an important role in many IR tasks.

Since search queries usually contain only a few words, solving the problem of data sparsity is particularly important for query reformulation understanding. Herdagdelen et al. present a novel approach utilizing syntactic and semantic information, whose operating costs are based on probabilistic term rewrite functions [6]. Dang et al. treat anchor text as a substitute for query log and study the effectiveness of a range of query reformulation techniques [5]. Sloan et al.

present a term-based methodology for interpreting query reformulation actions and make use of click data to test users' interaction behaviors when reformulating queries [3]. In this paper, instead of using the textual content of queries and results absolutely, we create query reformulation graph based on query logs and utilize translation module to help understand reformulation intentions.

2.2   Translating Embeddings

Existing related works aim to embed entities and relations of knowledge graphs into a continuous vector space, which provide great help to parse large graphs. These related works structure different translation functions to represent the relationship of head entities h, tail entities t and their relations r. **TransE** [10] interprets relations as translation operating from head entities to tail entities, which requires that t should be the nearest neighbor of $h+r$ in the vector space. **TransH** [11] and **TransR** [12] improve loss functions based on TransE by modeling each relation as translation on a hyperplane and a relation space respectively. **TransG** [13] proposes a generative model to address the issue of multiple relation semantics and **TransD** [14] also considers the problem of multiple type entities. **TransA** [15] utilizes an adaptive metric approach to provide a more flexible embedding method and another **TransA** [16] aims to find the optimal loss function by adaptively determining its margin over different knowledge graphs. **TranSparse** [17] deal with the heterogeneity and the imbalance of knowledge graphs, which are ignored by most previous works. In this paper, unlike conventional knowledge graph embedding methods, we propose a new framework to solve the issue of multilevel intentions in shopping query logs, which utilizes two-stage training method to calculate embedding result.

3   Methodology

3.1   Overall Framework

The framework of our method contains two modules, which shown in Fig. 2. We utilize translation embedding methods to train feature sets of relations and use classifiers to predict the best relation.

In translation module, the labeled data is split into two-level relations datasets, which contain all query pairs and only one level relations. These two datasets are inputted into translation embedding models separately. After algorithm training, we will obtain vector representation of all queries in two-level relations.

In classification module, we get the vector of each relation by calculating the difference of next query and current query. Each dimension of the relation vector is a feature of classifiers. Finally, we predict the best relation by combining the prediction results of two classifiers.

It is worth noting that the training set of translation embedding methods and classifiers are the same. Besides, we utilize 10-fold cross-validation to avoid data bias when training and testing.

Table 1. The description of two-level relations for query reformulation intentions. Last two columns show the example of query reformulation, where the keywords of second-level relations are highlighted with boldface.

Relation	Level	Description	Example	
Increase	1st-level	Adding a second-level relation	T-shirt	T-shirt+V-neck
Decrease	1st-level	Removing a second-level relation	T-shirt+Nike	T-shirt
Replace	1st-level	Replacing a second-level relation	T-shirt+O-neck	T-shirt+V-neck
Design	2nd-level	Shape of products	T-shirt	T-shirt+**V-neck**
Item	2nd-level	Name of products	**T-shirt**+Nike	**shoes**+Nike
Style	2nd-level	Series of products	T-shirt	T-shirt+**preppy**+**chic**
Brand	2nd-level	Brand or trademark of products	T-shirt	T-shirt+**Nike**
Model	2nd-level	Model or type of products	iPhone+**6**	iPhone+**X**
Function	2nd-level	Efficacy or utility of products	T-shirt	T-shirt+**anti-radiation**
Material	2nd-level	Material or stuff of products	T-shirt	T-shirt+**purified**+**cotton**
Channel	2nd-level	Shopping source of products	T-shirt	T-shirt+**flagship**+**store**
Cost	2nd-level	Price of products	T-shirt+**\$9.9**	T-shirt
Size	2nd-level	Size or measurement of products	T-shirt+**XL**	T-shirt+**3XL**

3.2 Translation Embedding

Given a training set S of the triplets (qc, qn, r) composed of current query $qc \in Q$ (the set of queries), next query $qn \in Q$ and labeled relationship $r \in R$ (the set of query reformulation intentions), the object of this module is to learn the vector embeddings of the queries. According to the definition of query reformulation intentions, **qn** should be a nearest neighbor of **qc** + **r** in the correct triples (qc, qn, r). The score function is defined as

$$f_r(qc, qn) = ||\mathbf{qc}_r + \mathbf{r} - \mathbf{qn}_r||_2^2 \tag{1}$$

In our model, we use **TransE**, **TransH** and **TransR** for achieving translation module. The meanings of $\mathbf{qc}_r$ and $\mathbf{qn}_r$ in these three models are different, where queries are mapped into a relation hyperplane in TransH and a relation space in TransR. We enforce constraints on the norms of the embeddings qc, r and qn uniformly, such as $||\mathbf{qc}||_2 \leq 1$, $||\mathbf{r}||_2 \leq 1$, $||\mathbf{qn}||_2 \leq 1$, $||\mathbf{qc}_r||_2 \leq 1$ and $||\mathbf{qn}_r||_2 \leq 1$.

To discriminate correct triplets and incorrect triplets, the margin-based loss function for training can be defined as

$$L = \sum_{(qc,r,qn)\in S} \sum_{(qc',r,qn')\in S'} max(0, d) \tag{2}$$
$$d = f_r(qc, qn) + \gamma - f_r(qc', qn')$$

where S is the set of correct triples, S' is the set of incorrect triples and γ is the margin.

Our labeled query reformulation logs only contain correct triples, so we construct incorrect triples $(qc', r, qn') \in S'$ by corrupting correct triples $(qc, r, qn) \in S$. We follow prior works and adopt stochastic gradient descent (SGD) to minimize the loss function.

Table 2. The percentage of the sample size of 30 relations in experimental dataset.

Relation	Design	Item	Style	Brand	Model	Function	Material	Channel	Cost	Size
Increase	34.0%	8.97%	7.47%	1.08%	0.77%	2.51%	1.74%	1.14%	0.17%	0.14%
Decrease	3.45%	0.70%	0.36%	0.31%	0.30%	0.30%	0.21%	0.34%	0.05%	0.04%
Replace	8.50%	20.6%	0.37%	4.23%	1.93%	0.08%	0.16%	0.01%	0.02%	0.02%
Sum	46.0%	30.3%	8.20%	5.62%	3.00%	2.88%	2.11%	1.49%	0.23%	0.20%

3.3 Classification

In this paper, we utilize several classifiers to accomplish classification module as follows.

- Random Forest Classifier (**RF**): An ensemble learning method for classification, that operates by constructing a multitude of decision trees at training time.
- Logistic Regression Classifier (**LR**): A regression model where the dependent variable is categorical.
- Gradient Boosting Decision Tree (**GBDT**): A prediction model in the form of an ensemble of weak prediction models.
- Support Vector Machine (**SVM**): A supervised learning model with associated learning algorithms that analyze data used for classification.
- K-Nearest Neighbor Classifier (**KNN**): A type of lazy learning, where the function is only approximated locally and all computation is deferred until classification.

The input data of classifiers include relation vectors and relation labels. Relation vectors are calculated by the difference of next query vectors and current query vectors from translation module, while relation labels are annotated by annotators. We use the accuracy of classifiers for evaluating the performance of our model, which corresponds to the evaluation metric of intentions prediction.

4 Experiments

In order to investigate the effectiveness of our method on modeling query reformulation, we compare the performance of our model to several baseline methods on intention prediction with a shopping search log dataset.

4.1 Dataset

We generate a new query reformulation dataset to evaluate our model. We collect 1.2 million users' search query logs on a day from a well-known shopping website, which contains 2.7 million search sessions. Those sessions containing only one query will be removed, because they are lack of query reformulation intentions. We build a directed graph using search query logs, where nodes represent queries and edges represent reformulation intentions. To make a small dataset, we select the subset of nodes with more than 50° and edges occurring more than 5 times. And then, we obtain a moderate dataset for labeling, which includes 19,335 query pairs and 12,085 queries.

We define two-level relations to represent query reformulation intentions, whose description is shown in Table 1. There are 3 first-level relations and 10 second-level relations, which can be combined into 30 relations. The intention of each query pair will be annotated manually by three annotators separately. Annotators may select multiple relations when one relation can't represent the intention completely. If a query pair doesn't match these 30 relations, annotators can annotate it as 'others'. According to the result of annotation fitting, there are 95.5% query pairs conform to these two-level relations, which prove the rationality of definition. We remove the multiple relations whose occurrence number is less than 100, and split query pairs with multiple relations into several single relations. After that, we get the labeled dataset with 18,630 query pairs and 11,331 queries, whose detailed statistical results in each relation is shown in Table 2.

Obviously, the distribution of dataset is uneven in different relations, which reflects the characteristics of shopping search. To evaluate experimental results of most query pairs actually, we remove the relation data of 'Channel', 'Cost' and 'Size', whose percentage of the sample size is less than 2%. Finally, we obtain the experimental dataset with 18,272 query pairs, 11,054 queries and 21 relations.

4.2 Baselines

Since we use **TransE**, **TransH** and **TransR** for carrying out translation module, we select these three models as baselines, which show outstanding performance in analyzing various knowledge graphs. Baselines don't contain a classification module, and their training and testing data don't divide into two-level relations. Referring to conventional knowledge graph embedding methods, we adopt $Hit@1$ as metric to evaluate the performance of relations prediction, which equates to the accuracy of our model mentioned above.

4.3 Overall Results

As shown in Table 3, we utilize accuracy of relations prediction to evaluate experimental results. The first row represents the result of three baselines without classification module and two-stage training, while the other rows represent the result of our model. From the results, we can observe that:

(1) Our proposed method consistently outperforms all baselines on all different combination of translation module and classification module, where the combinations containing 'KNN' show an increase of 20.6% of the average prediction accuracy. The reason is that the query logs of vertical search, such as shopping search, contain a large number of multilevel reformulation intentions, which can be solved well by two-stage training in our model. Since

Table 3. The overall accuracy of our model in relations prediction. The first row represents the results of three baselines without classification module, while the other rows represent the results of our model in the combination of different modules. Best result is highlighted with boldface.

Module	TransE	TransH	TransR
–	0.221	0.213	0.173
RF	0.263	0.262	0.255
LR	0.320	0.327	0.315
GBDT	0.331	0.331	0.325
SVM	0.347	0.354	0.351
KNN	0.407	**0.412**	0.407

the evaluation metric $Hit@1$ of conventional knowledge graph embedding methods don't perform well, we add classification module into our model and effectively improve the accuracy of prediction.

(2) In our model, the accuracy of the same classification module and different translation modules are similar, while the results of the same translation module and different classification modules differ a lot, where the combinations containing 'KNN' perform best. For query reformulation intentions predicting, we adopt accuracy instead of $Hit@k$ and $Mean$ used in knowledge graph embedding methods, because it is meaningful to predict users' reformulation intentions accurately. Obviously, the use of classifiers makes a great improvement, and different classifiers have different gains in the performance of our model.

Table 4. The prediction accuracy of first-level relations. The first row represents the results of three baselines without classification module, while the other rows represent the results of our model in the combination of different modules. Best result in each relation is highlighted with boldface.

Relation	Increase			Decrease			Replace		
Module	TransE	TransH	TransR	TransE	TransH	TransR	TransE	TransH	TransR
–	0.309	0.293	0.211	**0.305**	0.251	0.271	0.111	0.119	0.117
RF	0.225	0.226	0.213	0.093	0.093	0.095	0.331	0.327	0.327
LR	0.355	0.370	0.336	0.013	0.013	0.010	0.329	0.326	0.337
GBDT	0.225	0.229	0.220	0.153	0.162	0.156	0.474	0.471	0.468
SVM	**0.492**	0.484	0.478	0	0	0	0.239	0.263	0.264
KNN	0.329	0.334	0.337	0.198	0.203	0.230	0.526	**0.530**	0.511

4.4   First-Level Relations Prediction

Since different relations have different characteristics, such as the reflexive of 'Replace', we would like to measure the effectiveness of prediction in three first-level relations. In Table 4, we show the prediction accuracy of our model comparing baselines. It is worth noting that the test sample size of different relations is unequal, where 'Increase' is 6,398, 'Decrease' is 874 and 'Replace' is 5,773. From the table, we have the following observations:

(1) There is a great improvement in the results of 'Replace' for all combinations of our model. And in the results of 'Increase', our model also performs better than baselines except the combinations containing 'RF' or 'GBDT'. However, when testing the accuracy of 'Decrease', our model is lower than baselines. The reason is that the predictions of classifiers will tend to be the categories with a larger quantity. The test sample size of 'Decrease' is too small to be accurately predicted by the classifiers.
(2) In baselines, the results of 'Increase' and 'Decrease' have a good performance, while the effectiveness of 'Replace' is poor because of its reflexive attribute. Though TransE and TransR attempt to solve the issue of reflexive relations, it seems that the accuracy has not been greatly promoted in this shopping query dataset. To solve this problem, our model appends classification module and achieves significant improvement comparing to the baselines, where the accuracy of the combinations containing 'KNN' is over 50%.
(3) Comparing the results of different classifiers in our model, it is obvious that 'KNN' has the best overall effectiveness, which achieves advanced results in three first-level relations. And more importantly, there is not much difference of accuracy between 'KNN' and baselines when predicting 'Decrease', while other classifiers are far behind in performance. Thus, we select 'KNN' as the classification module in our model, and measure the effectiveness of second-level relations and top relations in the following sections.

4.5   Second-Level Relations Prediction

Compared with the first-level relations, the second-level relations can better reflect the needs of users for the product attributes. We calculate the prediction accuracy of second-level relations, which is shown in Table 5. The first column results represent the test sample size of 7 second-level relations, while the other columns results show the accuracy of three baselines and the improvement of 'KNN' module. Note that the sum of the test sample size of 7 second-level relations is smaller than training sample size, because some queries don't exist at the same time in the training set and test set when conducting translation module. From the results, we can find that:

(1) For the results of baselines, there is a great difference in the effectiveness of different relations. The accuracy of 'Model' is over 50%, while the accuracy of 'Item' is less than 10%. And the prediction results are not directly related to

Table 5. The prediction accuracy of second-level relations. The first column represents the test sample size of 7 second-level relations, while the other columns represent the results of three baselines. Improvement of accuracy added 'KNN' module based on baselines is shown in parentheses.

Relation	Count	TransE	TransH	TransR
Design	5,778	0.317 (+0.100)	0.258 (+0.165)	0.194 (+0.239)
Item	4,301	0.065 (+0.403)	0.081 (+0.390)	0.065 (+0.387)
Style	1,170	0.241 (−0.182)	0.344 (−0.291)	0.294 (−0.242)
Brand	778	0.240 (+0.427)	0.293 (+0.387)	0.225 (+0.429)
Model	450	0.542 (+0.054)	0.493 (+0.096)	0.542 (+0.011)
Function	318	0.094 (−0.037)	0.151 (−0.072)	0.164 (−0.098)
Material	250	0.140 (−0.072)	0.176 (−0.088)	0.180 (−0.108)

the test sample size. We judge that this phenomenon has a great relationship with the attributes of the second-level relations, where the definitions of some relations are clearer, while the others are ambiguous.

(2) Our model outperforms baselines in most second-level relations. In the results of 'Model', 'Function' and 'Material', the performance of our model and baselines is very close. And in the results of 'Design', 'Item' and 'Brand', our model has been significantly improved comparing baselines, where the accuracy of all these three relations is up to 40%. It is worth noting that the accuracy of 'Item' is raised from 7% to 46% averagely, which is the most obvious promotion. What's more, 'Item' is a top relation according to the quantity distribution of dataset in various relations. Therefore, this promotion will bring important practical significance for shopping query reformulation understanding.

(3) Our model is deficient in predicting the relations of 'style', where the accuracy is reduced from 29% to 5% averagely. This means that our model is hard to predict this relation, which often appears in the query logs about clothing. In order to understand the cause of this issue, we will give a detailed analysis in the section of case study.

4.6   Top Relations Prediction

There are some relations that often occur in the shopping search. Thus, the prediction accuracy of these relations is extremely important. We select 7 top relations, whose sample size is over 3%, and calculate the prediction accuracy for baselines and our model. The statistical results are shown in Table 6. The first column represents the test sample size of 7 top relations, while the other columns show the accuracy of three baselines and the improvement of 'KNN' module. From the results, we see that:

Table 6. The prediction accuracy of top relations. The first column represents the test sample size of 7 top relations, while the other columns represent the results of three baselines. Improvement of accuracy added 'KNN' module based on baselines is shown in parentheses.

Relation	Count	TransE	TransH	TransR
Increase-Design	3,814	0.383 (+0.072)	0.318 (+0.144)	0.191 (+0.280)
Replace-Item	3,315	0.037 (+0.506)	0.054 (+0.491)	0.036 (+0.488)
Replace-Design	1,406	0.142 (+0.251)	0.115 (+0.283)	0.176 (+0.222)
Increase-Style	1,064	0.256 (−0.197)	0.357 (−0.309)	0.309 (−0.255)
Increase-Item	888	0.152 (+0.062)	0.163 (+0.058)	0.155 (+0.059)
Replace-Brand	669	0.235 (+0.479)	0.287 (+0.450)	0.203 (+0.500)
Decrease-Design	558	0.308 (−0.089)	0.208 (+0.014)	0.260 (+0.003)

(1) The sum of the test sample size of 7 top relations is up to 90% of all test set, which means that the accuracy of these top relations will directly affect the overall performance. As mentioned above, we don't make the sample size of different relations equal by data sampling, because the uneven distribution of dataset is the characteristic of shopping search. In the same way, we don't use the average accuracy of different relations to evaluate the overall performance of our model and baselines. We think that the performance improvement brought by this evaluation metric will meet the actual needs of shopping search.

(2) Our model consistently achieves improvement comparing to all the baselines on all top relations except 'Increase-Style' and 'Decrease-Design'. Besides, the accuracy of top 3 relations is all up to 40%, where the sum of test sample size of these three relations is over 65% of all test set. Unlike conventional classifiers, whose classification results tend to the categories with larger training set, 'KNN' can perform very good effect in the category of small-scale training data, such as 'Replace-Brand'. Generally speaking, our model has been improved effectively based on baselines, and our model can accurately understand the query reformulation intentions of users.

(3) As the results of the previous section, our model suffers from the prediction of 'Increase-Style'. Because the test sample size of 'Style' and 'Increase-Style' is almost equal, we think that the reason for the low accuracy of these two relations is the same. We will carry out a detailed study on this issue in the following section.

4.7   Case Study

Our model suffers from low accuracy in predicting 'Style' as mentioned above. To explain the cause of this issue, we show the prediction results distribution of 'Style' in Table 7. Note that, for a test sample whose true category is 'Style',

Table 7. Prediction results distribution of second-level relation 'Style'. The decimal number represents the proportion of the predicted result to the test set of 'Style'.

Module	TransE			TransH			TransR		
Relation	Design	Item	Style	Design	Item	Style	Design	Item	Style
RF	0.888	0.070	0.029	0.909	0.064	0.020	0.879	0.074	0.026
LR	0.862	0.096	0.041	0.849	0.101	0.048	0.843	0.096	0.056
GBDT	0.868	0.061	0.050	0.894	0.048	0.035	0.879	0.050	0.045
SVM	0.952	0.048	0	0.944	0.056	0	0.949	0.051	0
KNN	0.879	0.036	0.075	0.903	0.029	0.062	0.89	0.039	0.063

the prediction result may be one of 7 second-level relations. We only show the relations with top 3 proportion, because the results of other relations are almost zero.

From the table, we find that about 90% of the test samples of 'Style' are predicted to be 'Design', which result in a very low prediction accuracy of 'Style'. The 'Design' describes the shape of the products, while the 'Style' represents the series of products. These two relations are intersecting when describing the products. What's more, there are 966 test samples containing 'Design' and 'Style' at the same time, 83% of the test samples of 'Style', which makes our model can't distinguish between these two relations. To solve this issue, we will redefine these two relations and deal with multiple relations better in the future.

5   Conclusion and Future Work

In this paper, we propose a novel method to model query reformulation for understanding users' intentions better. Specifically, we utilize translation module and classification module to predict relations in our framework, and propose two-stage training to improve the performance of prediction. We construct a new corpus of shopping search for experiments, and use accuracy as metric to evaluate the effectiveness of our model comparing baselines. Experimental results demonstrate that our method is effective for modeling query reformulation, where the prediction accuracy of many top relations is over 40%.

We will explore the following directions in future: (1) In shopping search, query reformulation can be seen as word set reformulation, whose semantic information can't be ignored. In future, we will add the semantic information of the query to our model, such as using RNN model to learn the initial vector of queries for translation module. (2) Users may not be satisfied with the search results of reformulation queries, which will affect our understanding of users' actual intentions. In future works, we will judge the satisfaction by the other users' behavior information, such as dwell time and click information, and use this satisfaction as the weight of each triple for our model.

References

1. Liu, Y., Wang, C., Zhou, K., Nie, J., Zhang, M., Ma, S.: From skimming to reading: a two-stage examination model for web search. In: ACM International Conference on Information and Knowledge Management, pp. 849–858. ACM (2014)
2. Huang, J., Efthimiadis, E.N.: Analyzing and evaluating query reformulation strategies in web search logs. In: Proceedings of the 18th ACM Conference on Information and Knowledge Management, pp. 77–86. ACM, November 2009
3. Sloan, M., Yang, H., Wang, J.: A term-based methodology for query reformulation understanding. Inf. Retr. J. **18**(2), 145–165 (2015)
4. Jiang, J.Y., Ke, Y.Y., Chien, P.Y., Cheng, P.J.: Learning user reformulation behavior for query auto-completion. In: Proceedings of the 37th International ACM SIGIR Conference on Research Development in Information Retrieval, pp. 445–454. ACM, July 2014
5. Dang, V., Croft, B.W.: Query reformulation using anchor text. In: Proceedings of the Third ACM International Conference on Web Search and Data Mining, pp. 41–50. ACM, February 2010
6. Herdagdelen, A., et al.: Generalized syntactic and semantic models of query reformulation. In: Proceedings of the 33rd International ACM SIGIR Conference on Research and Development in Information Retrieval, pp. 283–290. ACM, July 2010
7. Santos, R.L., Macdonald, C., Ounis, I.: Exploiting query reformulations for web search result diversification. In: Proceedings of the 19th International Conference on World Wide Web, pp. 881–890. ACM, April 2010
8. Hassan, A., Shi, X., Craswell, N., Ramsey, B.: Beyond clicks: query reformulation as a predictor of search satisfaction. In: Proceedings of the 22nd ACM International Conference on Information and Knowledge Management, pp. 2019–2028. ACM, October 2013
9. Bouramoul, A., Kholladi, M.K., Doan, B.L.: PRESY: a context based query reformulation tool for information retrieval on the web. arXiv preprint arXiv:1106.2289 (2011)
10. Bordes, A., Usunier, N., Garcia-Duran, A., Weston, J., Yakhnenko, O.: Translating embeddings for modeling multi-relational data. In: Advances in Neural Information Processing Systems, pp. 2787–2795 (2013)
11. Wang, Z., Zhang, J., Feng, J., Chen, Z.: Knowledge Graph Embedding by Translating on Hyperplanes. In: AAAI, vol. 14, pp. 1112–1119, July 2014
12. Lin, Y., Liu, Z., Sun, M., Liu, Y., Zhu, X.: Learning entity and relation embeddings for knowledge graph completion. In: AAAI, vol. 15, pp. 2181–2187, January 2015
13. Xiao, H., Huang, M., Hao, Y., Zhu, X.: TransG: a generative mixture model for knowledge graph embedding. arXiv preprint arXiv:1509.05488 (2015)
14. Ji, G., He, S., Xu, L., Liu, K., Zhao, J.: Knowledge graph embedding via dynamic mapping matrix. In: Proceedings of the 53rd Annual Meeting of the Association for Computational Linguistics and the 7th International Joint Conference on Natural Language Processing (Volume 1: Long Papers), vol. 1, pp. 687–696 (2015)
15. Xiao, H., Huang, M., Hao, Y., Zhu, X.: TransA: an adaptive approach for knowledge graph embedding. arXiv preprint arXiv:1509.05490 (2015)
16. Jia, Y., Wang, Y., Lin, H., Jin, X., Cheng, X.: Locally adaptive translation for knowledge graph embedding. In: AAAI, pp. 992–998, February 2016
17. Ji, G., Liu, K., He, S., Zhao, J.: Knowledge graph completion with adaptive sparse transfer matrix. In: AAAI, pp. 985–991, February 2016

A Deep Top-K Relevance Matching
Model for Ad-hoc Retrieval

Zhou Yang[1], Qingfeng Lan[2], Jiafeng Guo[3], Yixing Fan[3], Xiaofei Zhu[1(✉)],
Yanyan Lan[3], Yue Wang[1], and Xueqi Cheng[3]

[1] School of Computer Science and Engineering, Chongqing University of Technology,
Chongqing, China
yangzhou@software.ict.ac.cn, {zxf,wangyue}@cqut.edu.cn
[2] University of Chinese Academy of Sciences, Beijing, China
lanqingfeng14@mails.ucas.ac.cn
[3] CAS Key Lab of Network Data Science and Technology,
Institute of Computing Technology, Chinese Academy of Sciences,
Beijing 100190, China
{guojiafeng,lanyanyan}@ict.ac.cn, fanyixing@software.ict.ac.cn

Abstract. In this paper, we propose a novel model named DTMM,
which is specifically designed for ad-hoc retrieval. Given a query and a
document, DTMM firstly builds an word-level interaction matrix based
on word embeddings from query and document. At the same time, we
also compress the embeddings of both document word and query word
into a small dimension, to learn the importance of each word. Specifically, the compressed query word embedding is projected into the term
gating network, and the compressed document word embedding is concatenated into the interaction matrix. Then, we apply the top-k pooling
layer (i.e., ordered k-max pooling) on the interaction matrix, and get the
essential top relevance signals. The top relevance signals is associated
with each query term, and projected into a multi-layer perceptron neural network to obtain the query term level matching score. Finally, the
query term level matching scores are aggregated with the term gating
network to produce the final relevance score. We have tested our model
on two representative benchmark datasets. Experimental results show
that our model can significantly outperform existing baseline models.

Keywords: Deep learning · Relevance matching · Ad-hoc retrieval

1 Introduction

In traditional information retrieval models, the relevance of the document is
measured according to the exact matching signals. That is to say, the relevance
score is determined by the frequencies of query word from the document. These
models often face the typical term mismatch problem [1] since the semantic
matching signals are ignored. Recently, deep neural network have achieved great
success in many natural language processing tasks. At the same time, these deep

© Springer Nature Switzerland AG 2018
S. Zhang et al. (Eds.): CCIR 2018, LNCS 11168, pp. 16–27, 2018.
https://doi.org/10.1007/978-3-030-01012-6_2

neural networks have also been applied in information retrieval, which called neural information retrieval (i.e., NeuIR). Neural information retrieval models, which measure the relevance based on continuous embedding [6], have achieved significant progress.

It is of great importance to model the word importance in retrieval models. In traditional retrieval models, they measured the word importance based on the inverse document frequency (i.e., IDF), which have been a specific requirement of retrieval models [2]. As these models only considered document words which equal to the query word, thus, it is sufficient to only take IDF of query words into consideration. Recently, neural retrieval models employ deep neural network to model the semantic matching between query words and document words. In this way, those words which relate to the query words are also used to measure the relevance. However, existing neural retrieval models ignored the importance of these non-query words, which is also critical in relevance judgement. Take the following case as an example:

Query: Introduction of animals living in water, such as sharks
A fragment of document A: Dolphins swimming in the water are looking for food.
A fragment of document B: A yellow puppy fell into the water.

From the above example we can see, compared with the exact matching signal water, dolphins and puppy as similar matching signal appear in the document A, B respectively. Given the semantic environment provided by water and sharks in query, the importance of dolphins should be greater than puppy [7]. So, matching errors can easily occur without emphasizing the importance of document words. When the importance of words is emphasized, it will have a beneficial effect on correct matching.

In this work, we take the document word importance into consideration while modeling the relevance between query and document. Specifically, we proposed a deep top-k relevance matching model (DTMM) for ad-hoc retrieval. DTMM is a deep retrieval model which takes the raw text of query and document as input, and extract relevance signals automatically through deep neural network to produce the final relevance score. Specifically, DTMM firstly build an interaction matrix, where each element denotes the interaction between the corresponding query word and document word. At the same time, we compressed the embedding of document word into a small dimension, and concatenate into the interaction matrix. In this way, the interaction matrix can not only capture the matching signals but also the document importance. Then, we apply the top-k pooling layer on the interaction matrix, and obtain the essential top relevance signals. The top relevance signals is associated with each query term, and projected into a multi-layer perceptron neural network to get the query term level matching score. Finally, the query term level matching scores are aggregated with the term gating network to produce the final relevance score.

We have conducted extensive experiments to verify the effectiveness of our model. Specifically, we tested our model in two benchmark datasets (i.e., MQ2007 and Robust04). The experimental result show that DTMM can significantly outperform existing baselines on both datasets. For example, the improvement of DTMM over the best neural ranking model (i.e., DRMM) on MQ2007 is about 8% in terms of MAP metric.

The next section discusses related work. Section 3 presents the Deep Top-K Relevance Matching. Experimental methodology is discussed in Sect. 4 and evaluation results are presented in Sect. 5. Section 6 concludes.

2   Related Work

In matching task, most of the traditional models are based on the exact matching signals, such as BM25 [11] and query likelihood [13]. The advantage of this model is that it takes into account the exact matching signals and different matching requirements.

In current neural deep model for matching task, there are two types of models. One is representation-focused deep matching models, another is interaction-focused deep matching models. Due to their emphasis on matching, interaction-focused matching deep models is more suitable for IR tasks.

The representation-focused deep matching models, will lose the exact matching signals, and which is important for IR. In these models, they mainly learn to express the lower dimensions of queries and documents before interacting. In these learning process, the exact signals will be lost. For example, ACR-I [4] is a typical model of it, with distributed representations. Also famous are representation-focused models DSSM [5] and CDSSM [12] using letter-tri-gram.

In the interaction-focused deep matching models, words is represented by a low dimension vector, usually expressed by word embedding. The model produces an interaction matrix and then matches the documents by learning the information from the interaction matrix. The advantage of these models is that they make full use of the exact signals. MatchPyramid [11] is a CNN based model that is easy to implement and pays attention to the matching of word level, phrase level, sentence level, but ignores the diverse matching requirement. DRMM is an interaction-focused deep matching model, which gives special importance to three factors of relevance matching [2], but neglects the importance of document when using histograms to solve diverse matching requirement. In DeepRank [9], it imitates human retrieval process and achieves good results, but this model is complex and can not be parallelized. Interaction-focused deep matching models and representation-focused deep matching models address the ranking task problem from different perspectives, and can be combined in the future [8].

3   A Deep Top-K Relevance Matching Model

Based on the above analysis, in view of the existing problems in the existing model, We proposed DTMM to solve the problem of ignoring document words importance. Our model is based on the DRMM. In general, it can be divided into four parts. The first part is to build the interaction matrix. DTMM constructs an interaction matrix with distributed representations of query and document words, In order to emphasize the importance of document, it adds the weight of document into the current interaction matrix to form a new interaction matrix. The second part is the k-max pooling layer, which selects the top k strongest signals with the query dimension as the input of the next layer. The third part is fully connected network. The model sends the information into a multi-layer neural network and combines the weight value of the term gating network with the query terms to get a final score. After this, it uses hinge loss function as the objective function.

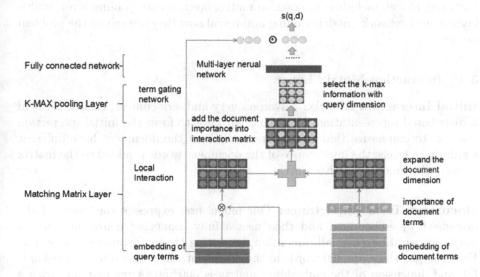

Fig. 1. Architecture of deep Top-k relevance matching model.

Formally, suppose that query and document are given in the form of vector sets: $q = \{q_1, q_2, \ldots q_M\}$ and $d = \{d_1, d_2, \ldots d_N\}$ represent the embedding of query and document separately, where M denotes the length of query, N denotes the length of document. The models we propose are as follows:

$$g_{qi} = softmax(w_{qi}q_i), g_{dj} = w_{dj}d_j \quad i = 1, 2, \ldots M, \quad j = 1, 2, \ldots N \quad (1)$$

$$z_i^0 = T_k(q_i \cdot d + g_{dj}) \quad i = 1, 2, \ldots M, \quad j = 1, 2, \ldots N \quad (2)$$

$$z_i^k = a_k(w^k z_i^{k-1} + b^k) \quad i = 1, 2, \ldots M, \quad j = 1, 2, \ldots N \quad (3)$$

$$s = \sum_{i=1}^{M} g_{qi} z_i^L \quad i = 1, 2, \ldots M \quad (4)$$

In detail, g_{qi} and g_{dj} represent the weight of query and document words respectively, and w_{qi} and w_{dj} are the weights of corresponding neural nodes respectively. The difference is that the importance of query words is calculated from softmax. $\cdot$ denotes the interaction operator between a query term and the document term, and DTMM model uses dot as interaction operator. T_k is a Top-K pooling function, namely k-max pooling function with the query dimension. z_i^k denotes the output of each neural layer, specially, z^0 is the first layer denoting interaction matrix. In the second equation, w^k and b^k are the weight matrix and bias of the k layer neural network, respectively. a_k is the activation function of each neural network layer, here we use softplus. g_i is the weight coefficient of the i-th query word, s is the final score of the model. The structure of the model is shown in Fig. 1.

Next we will discuss the main components in the deep Top-K relevance matching model, including interaction matrix layer, k-max pooling layer, multi-layer neural network, model training and reveal how they are solved the problem before.

3.1   Interaction Matrix Layer

Initial Interaction Matrix: Given a query and a document, each word is in a distributed representation, interacted with dot to form the initial interaction matrix. To emphasize that the different words in the document have different significance levels, the importance of the document words is added to the matrix to form the final interactive matrix.

Document Term Importance: Our model first expresses the words of the document by embedding, and then uses a fully connected neural network to map each word into a small dimension, indicating the importance of the words. For example, there are 300 words in the document, the embedding dimension is 50, and dimension of the embedding matrix is $300*50$. After mapping with a fully connected neural network, it is represented as $300*1$ dimension. We need to add document term importance into the interaction matrix. Concretely, if the query has 5 words, then expanding the document term importance matrix to $5*300*1$ dimension, then add it to the corresponding interaction matrix before.

3.2   K-max Pooling Layer

We observed that in essence, the matching histograms used in the DRMM model is actually sorted, in the meantime, the unimportant words in document are also added to the calculation of histograms. Through our research, A few words in document with a high correlation of the query basically determine the final score, while a large number of low correlation degrees are found. The low correlation signals, like stop words is little or has negative impact. On this basis, we propose a top-k pooling function based on query dimension to select the optimal signals, removing the bad signals. After processing of k-max pooling layer, the length of whole document dimension is K [3]. It forms a fixed value and provides the conditions for entering a fully connected neural network.

In addition, in order to achieve end to end training, we use the k-max pooling function instead of the histograms in the DRMM model. It make full use of the existing word embedding to accelerate the training, it breaks through the limitations of the original word embedding and maximally avoids disadvantages.

3.3 Multi-layer Neural Network

Query Term Importance: Since importance of different query words is different, we use a weight coefficient to distinguish it. The greater the weight have, the more important the word is in the query. DTMM use a weighted network to calculate the weight coefficients of different query words. Specifically, we use the softmax function as a activation function. All of this imitates query term importance in DRMM.

Multi-layer Neural Network: After that, DTMM constructs a multi-layer neural network. Based on the powerful data fitting and learning ability of neural network, the features are extracted one by one. With the gradual deepening of the network, the extracted features become more and more abstract. Because the importance of each query is different, the output of the multi-layer neural network combined with the query importance to get a final score, which is used for document sorting.

3.4 Model Training

The task of Ad-hoc information retrieval is a kind of sorting problem, so we use one of the sort loss functions hinge loss as the loss function in the model training.

$$L(\theta) = mean[\sum_{q} \sum_{d^+ \in D_q^+, d^- \in D_q^-} max(0, 1 - s(q, d^+) + s(q, d^-))] \qquad (5)$$

In detail, θ represents all the parameters to be learned in the model, q denotes query, d^+ comes from the positive sample document sets D^+, which represents the documents that is positively related to the query. d^- comes from the negative sample document sets D^-, which represents the documents that is not related to the query. The method widen the gap between positive and negative samples, and makes the positive score larger than the negative case over 1. DTMM is optimized by back-propagation algorithm.

4 Experimental Methodology

In this section, We compared with several classical models and achieved good results. Next, we elaborate on the detailed process, results and analysis of our experiments.

4.1 Dataset

Million Query Track 2007: It is called MQ2007 for short. The data set is a subset of the LETOR4.0, which is collected by the web crawler from the domain name GOV2 web site, and the user clicks are used as the basis for the sorting of the document, Including 25M documents and 10000 queries. MQ2007 has a total of 58730 documents and 1501 queries. Among them, the words in the document and the query are lowercased and indexed, and the corresponding words are extracted with the Krovetz stem analyzer. In addition, referring to the list of stop words in INQUERY, we removed the stop words in the query. The parameters of the data set are detailed in Table 1.

Robust04: Robust04 is a small news dataset. Here we use Robust04-title as one of our data set. The topics are collected from TREC Robust Track 2004. Here the Robust04-Title means that the title of the topic are used as query. The collection is consist of 0.5M documents and 250 queries. The vocabulary size is 0.6M, and the collection length is 252M. More clearly described in the lower Table 1.

Table 1. Statistics of collections used in this study. Here we tested our model DTMM on two data sets MQ2007 and robust04.

	MQ2007	robust04
Query number	1501	250
Document number	58730	324541

4.2 Baseline Methods

Our baseline includes traditional models, including BM25 [11], and some recent neural ranking models. One type is representation-focused deep matching models, including ACR-I [4], DSSM [5], CDSSM [12], and another interaction-focused deep matching models as follows: ACR-II [4], MatchPyramid [11], DRMM [2]. In detail, when we use the MQ2007 as our dataset, the metrics of all of our models are set in [9], and when we use robust04-title, the metrics we referencing to are set in [2].

We select some neural deep matching models for comparison, and we will introduce these models below:

ARC-I: ARC-I is used in sentence completion, response matching, and paraphrase identification, which is a representation-focused model. ARC-I has been tested on a set of NLP tasks including response matching, sentence completion, and paraphrase identification.

DSSM: DSSM is a excellent model for web search. The original paper mentioned that training DSSM requires a lot of data. In the following experiments, it does not show excellent results.

CDSSM: DSSM is an improved version of CDSSM. It mainly changes the dense layer in DSSM to the convolution layer, getting more structural information by this way, and has improvement in performance.

ARC-II: It is an improved version of ACI-I. It has noticed the importance of interaction, and has learned interactive information earlier than ARC-I. ARC-I and ARC-II has no public code, so it is re-implemented and applied to the comparison model [2].

MatchPyramid: It is a widely used model, and its applications include paraphrase identification and paper citation matching. There are three versions of MatchPyramid. We choose the best models to compare. The model involved in the comparison is the original model provided by the author.

DRMM: DRMM is an interaction-focused model, With different types of histogram mapping functions (i.e., CH, NH and LCH) and term gating functions (i.e., TV and IDF). We choose the best model of the result to compare. Similarly, the model involved in the comparison is the original model provided by the author.

4.3 Implementation Details

This section describes the configurations of DTMM.

Embedding Size: We use the embedding with 50 dimension size, it is trained by the GloVe [10] model in advance. During the training process, we have not synchronized training embedding due to the small amount of data. Through our statistics, the vocabulary of corpus is 193367.

K-max Pooling Layer Size: The k-max pooling layer selects 512 optimal signals, and other weak signals will not be input into the neural network. Through our research, different features and number of it in the data set affect the setting of this parameter size.

Multilayer Neural Network Size: The size of the multi-layer neural network is set to [512, 512, 256, 128, 64, 32, 16, 1], with activation function of softplus.

Model Optimization: Optimization using Adam optimizer, with e $= 1 - 5$, learning rate $= 0.001$ and batch size $= 100$. We conducted on MatchZoo development, it is an open source matching model development platform using keras tensorflow, including most advanced matching model nowadays.[1]

5   Evaluation Results

5.1   Ranking Accuracy

Obviously, our proposed DTMM model has a significant improvement over baseline. The experimental results of the model in MQ2007 and robust04 are as follows:

On the MQ2007 data set, all the representation-focused models (including DSSM, CDSSM, ARC-I) and most interaction-focused models (including DRMM, ARC-II, MatchPyramid) do not work as well as BM25 [11]. In the previous model, only DRMM performs better than the BM25. The performance of

---

[1] The source of MatchZoo: https://github.com/faneshion/MatchZoo.

Table 2. Comparison of different retrieval models over the MQ2007.

Model	NDCG@1	NDCG@3	NDCG@5	NDCG@10	P@1	P@3	P@5	P@10	MAP
BM25	0.358	0.372	0.384	0.414	0.427	0.404	0.388	0.366	0.450
DSSM	0.290	0.319	0.335	0.371	0.345	0.359	0.359	0.352	0.409
CDSSM	0.288	0.288	0.297	0.325	0.333	0.309	0.301	0.291	0.364
ARC-I	0.310	0.334	0.348	0.386	0.376	0.377	0.370	0.364	0.417
DRMM	0.380	0.396	0.408	0.440	0.450	0.430	0.417	0.388	0.467
ARC-II	0.317	0.338	0.354	0.390	0.379	0.378	0.377	0.366	0.421
MatchPyramid	0.362	0.364	0.379	0.409	0.428	0.404	0.397	0.371	0.434
DTMM	0.458	0.459	0.468	0.499	0.517	0.479	0.458	0.426	0.504

Table 3. Comparison of different retrieval models over the robust04.

Model	NDCG20	P@20	MAP
BM25	0.418	0.370	0.255
DSSM	0.201	0.171	0.095
CDSSM	0.146	0.125	0.067
ARC-I	0.066	0.065	0.041
DRMM	0.431	0.382	0.279
ARC-II	0.147	0.128	0.067
MatchPyramid	0.330	0.290	0.189
DTMM	0.463	0.432	0.314

the representation-focused models is generally not as good as the performance of interaction-focused model. To some extent, this illustrates the role of the three factors emphasized by relevance matching in IR. The improvement of DTMM against the best deep learning baseline (i.e. DRMM) on MQ2007 is 20.6% wrt NDCG@1, 15% wrt P@1, 8% wrt MAP, which illustrates the superiority of our model on the IR task (Tables 2 and 3).

On the robust04 data set, the performance of most interaction-focused models is also obviously better than the representation-focused models. But one exception is that the interaction-focused model ARC-II has the same performance as the CDSSM, and is inferior to the representation-focused model DSSM. This may be during to the uneven distribution of features in this data set. When ARC-II intercepts text length, it removes the important feature at the end of the document, which has an impact on model performance. In the same way, most of the interaction-focused models and representation-focused models can not exceed BM25 performance except DRMM model. On this data set, DTMM also achieves the best effect, compared to the best model DRMM. The improvement of DTMM against the best deep learning baseline (i.e. DRMM) on robust04 is 7.4% wrt NDCG@20, 13% wrt P@20, 12.5% wrt MAP, respectively. It needs to be explained that the reason we choose these indexes is because on the MQ2007 dataset, we refer to [2], and on the robust04 refer to, we refer to [9].

Table 4. Comparison of different version of DTMM. Where $DTMM_{no}$ represents the model without document words importance, the other is the complete model.

Model	NDCG@3	NDCG@5	NDCG@10	MAP
$DTMM_{no}$	0.424	0.435	0.469	0.490
DTMM	0.459	0.468	0.499	0.504

5.2 Performance on DTMM Without Document Words Importance

Table 4 shows the comparison between DTMM and DTMM without adding document word importance. Where $DTMM_{no}$ represents the model without document words importance, the other is the complete model. In the evaluation of ndcg@3, ndcg@5, ndcg@10 and MAP, the complete DTMM was higher than the incomplete model 8.25%, 7.58%, 6.39%, 2.85% respectively. It shows that emphasizes the importance of the different words in the document is meaningful.

5.3 Performance on Different K-max Pooling Layer of DTMM

Figure 2 shows the performance of different versions of DTMM on the MQ2007 data set. Obviously, with the parameter selection from small to large, the performance of the model first improves and then decreases. It can be seen that different top-k has an impact on the performance of the model. Since each data set has different characteristic types and numbers, it is important to select the parameter Top-k. When the selected Top-k is too large or too small, it will have

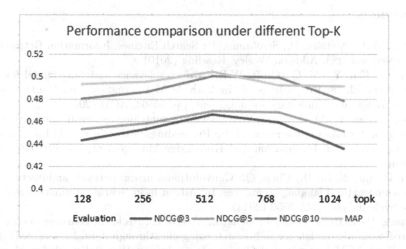

Fig. 2. Performance comparison on DTMM with different Top-K. The abscissa is the size of the top k window, and the ordinate is the size of the indicator. Our indicators include NDCG@3, NDCG@5, NDCG@10, MAP

different effects on the model. If it is too small, the selected features are insufficient. If the Top-k is too large, the unimportant information is selected and it will have an impact on the performance.

6   Conclusion

This article presents a DTMM model for ad-hoc tasks. It emphasizes the importance of document words for IR task, and also proposes a k-max pooling method based on query dimension, which can eliminate noise while retaining the strongest signals. The model includes three parts: interaction matrix layer, k-max pooling layer and multi-layer neural network. Each part of the model can be parallelized, making large-scale commercial offerings possible.

For the future work, as DTMM is at word-level, we will consider adding phrase-level and sentence-level matching signals to the it. More generally, we will delve into the factors that are favorable to IR and introduce them into the new model.

Acknowledgments. This work was funded by the 973 Program of China under Grant No. 2014CB340401, the National Natural Science Foundation of China (NSFC) under Grants No. 61425016, 61472401, 61722211, and 20180290, the Youth Innovation Promotion Association CAS under Grants No. 20144310, and 2016102, and the National Key R&D Program of China under Grants No. 2016QY02D0405. National Natural Science Foundation of China (No. 61603065, No. 61502064, No. 61702063), Foundation and Frontier Research Key Program of Chongqing Science and Technology Commission (Grant No. cstc2017jcyjBX0059, No. cstc2017jcyjAX0277, No. cstc2017jcyjAX0089).

References

1. Croft, W.B., Metzler, D., Strohman, T.: Search Engines: Information Retrieval in Practice, vol. 283. Addison-Wesley, Reading (2010)
2. Guo, J., Fan, Y., Ai, Q., Croft, W.B.: A deep relevance matching model for Ad-hoc retrieval. In: Proceedings of the 25th ACM International on Conference on Information and Knowledge Management, pp. 55–64. ACM (2016)
3. Guo, J., Fan, Y., Ai, Q., Croft, W.B.: Semantic matching by non-linear word transportation for information retrieval. In: Proceedings of the 25th ACM International on Conference on Information and Knowledge Management, pp. 701–710. ACM (2016)
4. Hu, B., Lu, Z., Li, H., Chen, Q.: Convolutional neural network architectures for matching natural language sentences. In: Advances in Neural Information Processing Systems, pp. 2042–2050 (2014)
5. Huang, P.-S., He, X., Gao, J., Deng, L., Acero, A., Heck, L.: Learning deep structured semantic models for web search using clickthrough data. In: Proceedings of the 22nd ACM International Conference on Information & Knowledge Management, pp. 2333–2338. ACM (2013)
6. Mikolov, T., Sutskever, I., Chen, K., Corrado, G.S., Dean, J.: Distributed representations of words and phrases and their compositionality. In: Advances in Neural Information Processing Systems, pp. 3111–3119 (2013)

7. Mikolov, T., Yih, W.-T., Zweig, G.: Linguistic regularities in continuous space word representations. In: Proceedings of the 2013 Conference of the North American Chapter of the Association for Computational Linguistics: Human Language Technologies, pp. 746–751 (2013)
8. Mitra, B., Diaz, F., Craswell, N.: Learning to match using local and distributed representations of text for web search. In: Proceedings of the 26th International Conference on World Wide Web, pp. 1291–1299. International World Wide Web Conferences Steering Committee (2017)
9. Pang, L., Lan, Y., Guo, J., Xu, J., Wan, S., Cheng, X.: Text matching as image recognition. In: AAAI, pp. 2793–2799 (2016)
10. Pennington, J., Socher, R., Manning, C.: GloVe: global vectors for word representation. In: Proceedings of the 2014 Conference on Empirical Methods in Natural Language Processing (EMNLP), pp. 1532–1543 (2014)
11. Robertson, S.E., Walker, S.: Some simple effective approximations to the 2-poisson model for probabilistic weighted retrieval. In: Croft, B.W., van Rijsbergen, C.J. (eds.) SIGIR '94, pp. 232–241. Springer, New York (1994). https://doi.org/10.1007/978-1-4471-2099-5_24
12. Shen, Y., He, X., Gao, J., Deng, L., Mesnil, G.: Learning semantic representations using convolutional neural networks for web search. In: Proceedings of the 23rd International Conference on World Wide Web, pp. 373–374. ACM (2014)
13. Zhai, C., Lafferty, J.: A study of smoothing methods for language models applied to information retrieval. ACM Trans. Inf. Syst. (TOIS) 22(2), 179–214 (2004)

A Comparison Between Term-Based and Embedding-Based Methods for Initial Retrieval

Tonglei Guo[✉], Jiafeng Guo, Yixing Fan, Yanyan Lan, Jun Xu, and Xueqi Cheng

CAS Key Lab of Network Data Science and Technology, Institute of Computing Technology, Chinese Academy of Sciences, Beijing 100190, China
{guotonglei,fanyixing}@software.ict.ac.cn,
{guojiafeng,lanyanyan,junxu,cxq}@ict.ac.cn

Abstract. The initial retrieval stage of information retrieval aims to generate as many relevant candidate documents as possible in a simple yet efficient way. Traditional term based retrieval methods like BM25 deal with the problem based on Bag-of-Words (BoW) representation, thus they only focus on exact matching (i.e., syntactic) and lack the consideration for semantically related words. That causes the typical vocabulary mismatch problem and the reduction of performance in terms of recall. The advance of distributed representation (i.e., embedding) of words and documents provides an efficient way to measure the semantic relevance between words. Since embedding can alleviate the vocabulary mismatch problem, it is suitable for the initial retrieval task. We conduct several experiments to compare term based models with embedding based models in terms of recall. We compare above two branches of the initial retrieval models on three representative retrieval tasks (Web-QA, Ad-hoc retrieval and CQA respectively). The results show that embedding based method and term based method are complementary for each other and higher recall can be achieved by combining the above two types of models based on scores or ranking position. We find that combination of the two types of the models based on ranking position usually perform better than combination based on score. Furthermore, since queries and documents are in different forms for diverse application scenarios, it can be observed that the relative performance of the two types are almost same but the absolute performance are significant different regarding to distinct scenarios.

Keywords: Initial retrieval · Embedding representation · Recall

1 Introduction

The process of information retrieval can typically divide into two stages, namely the initial retrieval stage and the re-ranking stage [5]. The initial retrieval stage aims to retrieve a small candidate set containing as many relevant documents as possible. It is required to achieve high recall in a simple yet efficient way. Then,

© Springer Nature Switzerland AG 2018
S. Zhang et al. (Eds.): CCIR 2018, LNCS 11168, pp. 28–40, 2018.
https://doi.org/10.1007/978-3-030-01012-6_3

the re-ranking stage applies complicated methods like various deep models to rank the candidate document according to their relevance with the given query. In this paper, we focus on the initial retrieval stage.

Traditional initial retrieval models use term based methods for high efficiency. Term based models are relied on Bag-of-Words (BoW) representation and assumes words in documents are independent from others, the measurement of relevance relies on exact matching (i.e., syntactic or terms counting) of words. But relevant texts sometimes can be expressed in totally different words, e.g., "How bad is the new book by J.K. Rowling?" and "How is the new Harry Potter book Harry Potter and the Cursed Child?" are related obviously but they share little common words. This causes the typical problem in IR called *vocabulary mismatch* and leads to many relevant documents failed to be retrieved in this stage.

The advance of distributed models [14, 17, 20] provides an efficient way to represent queries and documents semantically. These models can represent a single word or a piece of text by a vector in a continuous semantic vector space, referred to as "embedding". These embedding representations have shown promising results in NLP, which inspires the application in the initial retrieval stage to capture the semantic relevance.

In this paper, we explore term based models and embedding based models in the initial retrieval step. We choose BM25 and language model to represent for term based models. Existing embedding based models can be divided into aggregated distributed representation and paragraph vector representation. We use BoWE model and an enhanced model based on PV to represent two parts. Furthermore, we explore the hybrid models which combine two branches of models to take both exact matching and semantic relevance into account. We evaluate the recall of each model in three typical IR applications: Ad-hoc retrieval, Web-QA and community-based question answering (CQA). We give brief introduction of these applications and describe the characteristic of different scenarios. We design experiments on three benchmark datasets respectively representing three scenarios and do some analysis according to the results. The performance is measured in terms of recall concerned in the initial retrieval step.

In brief, here are some take-away conclusions based on our experimental results:

1. Considering the performance in terms of recall for each method independently, embedding based methods can't outperform term based methods, but it can be seen that cases where they perform well are not overlapped, in other words, they are complementary to each other.
2. Comparing the two embedding based methods, the BoWE has higher recall than the method of PV. There is little difference between the recall of two embedding based methods in CQA task. But when it comes to Ad-hoc retrieval and Web-QA, PV shows poor performance than BoWE.
3. Usually the combination of the term based methods with embedding based methods can outperform term based methods. Moreover, the hybrid of two branches of models based on ranking position usually do better than the combination based on score.

The rest of this paper is organized as follows: In Sect. 2, we discuss some related work regrading to the initial retrieval step. In Sect. 3, we describe the two branches of the initial retrieval model in details. Then we show our experimental results and do some analysis in Sect. 4.

2 Related Work

In this section, we overview the researches related to our works in three areas: term based models, embedding based models and combination models for the initial retrieval step.

2.1 Information Retrieval Process

Information retrieval can be typically divided into two stages [5], namely the initial retrieval stage and the re-ranking stage. Since modern IR system usually need to deal with a huge amount of data, it is impossible to rank all documents given a query. The initial retrieval stage aims to select a small subset of documents containing as many relevant documents as possible (i.e., high recall) in an efficient way. Traditional initial retrieval models are term based models like BM25 [23] since their simplicity and efficiency. Modern IR system achieved this step by building a symbolic based inverted index and a search scheme. On the other hand, the re-ranking stage focus on ranking the above candidate documents according to the relevance of the given query as precise as possible (i.e., high precision). Since the size of the subset is relative small and the requirement of precision, so more complicated models are used in re-ranking stage, for example, various learning to rank algorithms [2,3] and deep learning models [8,18].

2.2 Embedding

The revival and new development of neural networks in natural language processing (NLP) has brought the attention of using these techniques in information retrieval tasks. Existing works of applying neural networks in IR mainly consists of two branches, the first one is utilizing embeddings to represent query and document [9,25], the other one is using deep models to learn representations of query and document [11]. Most deep learning models are complicated so usually they are explored in the re-ranking stage. While the embeddings can be pre-trained and the relevance calculation is simplified to the vector similarity, so it is more promising in the initial retrieval stage.

The embeddings of words and documents in a low-dimensional vector space have attracted many researchers. The ability of embeddings which can encode semantic and syntactic relations between words and documents seems benefits retrieval models in semantic matching. Existing works generally utilize word2vec [17] to obtain pre-trained word embeddings and then utilize them in retrieval tasks [9,25]. Besides above methods which combines pre-trained word embeddings, there are some methods learn embeddings from scratch directly.

Le et al. [14] propose doc2vec which learns words and documents embeddings representations together. The learned representation can be used to measure the relevance between queries and documents by the similarity of embedding vectors.

3 Initial Retrieval Models

Term based models calculate the relevance based on the occur frequencies of query words, i.e., exact matching which causes mismatch of many semantic related word pairs, e.g., "metro" and "subway". That leads to the typical vocabulary mismatch problem in IR and degrades the performance of retrieval models. The recently proposed word embedding, which can capture semantic relatedness between words, have attracted a lot of attentions in many NLP tasks. The relevance can be calculated efficiently from vector similarity so it is worthwhile to explore the embedding based initial retrieval models.

In this section, we explore the embedding based models in the initial retrieval step and use them to enhance the term based models. We start with the term based models in Sect. 3.1 and then introduce two kinds of embeddings based models in detail. In Sect. 3.3, we describe the hybrid models of above models in two ways. At last, we give brief introduction of three typical IR applications where above models are used in.

3.1 Term Based Initial Retrieval Models

Most term based models for the initial retrieval step use BoW representation based on the assumption that words are independent from others. In this way, each word can be viewed as a dimension in a orthogonal semantic vector space where each query or document is represented as a point. A query and a document which is close to each other will be considered as relevant pairs.

BM25. BM25 [23] is a representative state-of-the-art retrieval model which based on the probabilistic ranking principle. Both query and document are represented as a vector using bag-of-words assumption. BM25 calculate the relevance score of the document based on the query term frequencies and document length: $BM25(q,d) = \sum_{i=1}^{n} IDF(q_i) \cdot \frac{f_i \cdot (k_1+1)}{f_i + k_1 \cdot (1 - b + b \cdot \frac{dl}{avgdl})}$. Where k and b are tunable hyper-parameters.

Language Model. Language model [22,26,27] is based on a probabilistic model which predict the probability that a query q is generated by a document d. According to the Bayes'formula, $p(d|q) \propto p(q|d) \cdot p(d)$. Thus the relevance of the given query q and a candidate document d is measured by the conditional probability $p(d|q)$. Language models usually have various smoothing methods [22,26,27] to overcome the zero probability of unseen words and data sparseness.

3.2   Embedding Based Initial Retrieval Models

Distributed representation, also named embedding, which is learned based on the distributional hypothesis [6,10], has achieved a lot of success in natural language processing. It has been found that the distributed representation of documents can well capture the semantic relatedness of documents [17]. Thus, it is straightforward to employ document embedding to build the retrieval model in the initial retrieval stage. Embedding based retrieval models can be categorized into two classes according to the way of constructing document embedding, namely aggregated distributional representation and paragraph vector representation. In the follows, we will describe each model in detail.

Aggregated Distributed Representation. Aggregated distributed representation views a word in a text as the basic unit and aggregated embeddings of these words in a specific way to obtain the final document representation. There have been a large number of works [9,25] which studied how to integrate the word embeddings to obtain the document representation in an effective way. In this paper, we directly utilized the simple but effective aggregated model which is based on weighted sum [9]. Denote D is the collection of all documents and d is a document in D. BoWE representation views a documents as a bag of words in d. Given the word embedding matrix $W \in R^{K \times |V|}$ for a finite size V of vocabulary consists of all words from the corpus. Each column $W^i \in R^k$ is the K dimensional word embedding of the i-th words. Then the document d can be expressed as $d = w_1^d, w_2^d, \ldots, w_n^d$, n is the number of words in d and w_i^d denotes the embedding of i-th words. We represent the document d as a fixed length vector in following way:

$$\vec{d} = IDF_1 \cdot \vec{w_1} + IDF_2 \cdot \vec{w_2} + \cdots + IDF_n \cdot \vec{w_n} \tag{1}$$

Paragraph Vector Representation. To utilize the information of words order, there comes another embeddings way which views a piece of paragraph as the basic unit [14]. In this paper we use an enhanced model based on the PV-DBOW model [14] which predict unigram, bigram and trigram words randomly sampled from the paragraph. Each n-gram of words is represented by a distinct embeddings.

3.3   Hybrid Models

Term based models lost semantic matching of words which limits the performance of models. To overcome the defect of vocabulary mismatch problem and consider semantic relevance, we propose two hybrid models to combine term based models with embedding based models. Let $f_{BoW}(q, d)$ denote a term based model and $f_{emb}(q, d)$ denote an embedding based model for measuring the relevance of a given query q and a candidate document d.

Score-Based Hybrid Model. This model utilize the score of two branches of models directly. It calculates the final score while linearly combining the score results of two branches models.

$$f_s(q, d)_{score} = (1 - \lambda) \cdot f_{BoW}(q, d)_{score} + \lambda \cdot f_{emb}(q, d)_{score} \tag{2}$$

Rank-Based Hybrid Model. This model is based on the result of ranking positions obtained by the two branches of models. It ignores the exact score value but focus on the positions in the ranking lists. The scores used for combination is the reciprocal of the position in the resulting ranking lists: $s(q, d) = \frac{1}{r}$.

$$f_s(q, d)_{rank} = (1 - \lambda) \cdot f_{BoW}(q, d)_{rank} + \lambda \cdot f_{emb}(q, d)_{rank} \tag{3}$$

3.4 Typical IR Applications

Information retrieval (IR) has various applications, such as Ad-hoc retrieval, Web-Question-Answering (Web-QA) and community question answering (CQA) etc. This paper focus on three typical IR applications, so in this section, we give brief review of these three IR applications.

Ad-hoc Retrieval. The core problem of Ad-hoc retrieval [11,16] is to measure the relevance for a document given the question. The task of ad-hoc retrieval is to fetch the relevant candidate documents and rerank them according to the relevance given a query. Usually the documents in the dataset are stable while the queries are various with different users. The initial retrieval step in ad-hoc retrieval is required to return as many related documents as possible and leave the re-ranking to the next step.

Web-QA. Web-QA [7,15,21] is based on IR models which relies on the massive data formalized as texts on the web. The task of web-QA is to use information retrieval techniques to extract the relevant documents on web given a specific question. A web-QA system would firstly process the question in natural language to understand the question properly like the answer type etc. and then, formulates a query for a search engine to retrieve relevant documents. Finally the correct answer would be extracted from the most related documents. For the initial retrieval step, the web-QA models only focus on retrieve relevant documents which might contain correct answer according to the question.

CQA. Community question answering (CQA) [1,4,19] is increased in popularity with the advance of many community forums. Its task is to find the correct answer of the given question from the dataset contributed by all the users. Different from the web-QA system which tries to find answer directly, a CQA system would check if there is a duplicated question and return the answer of this question. If there isn't any similar question with the given one, then the system leave it to other users to answer the question and allow them rate the answers. In this paper, we focus on the initial retrieval step which tries to find similar questions with the given question.

4 Experiement

In this section, we conduct experiments in three typical IR application: Ad-hoc retrieval, Web-QA and community-based question answering(CQA). We choose

Table 1. The feature of three typical IR applications.

Application	Text1	Text2	Type
Ad-hoc retrieval	"International organized crime"	"Guatemalan President has sent a letter to U.S. Secretary of State Warren Christopher to promote the visit of U.S. tourists to Guatemala, saying the recent acts of violence against two U.S. citizens were isolated incidents. Due to those acts of violence, the U.S. Government has issued a travel warning advising travelers not to visit Guatemala or to postpone their trips to that country, except for emergency purposes."	Short-long
Web-QA	"How to remove tree sap from car"	"I have had good luck with commercial products available at auto supply stores for just a few dollars below are some additional suggestions if you want to try DIY"	Medium-medium
CQA	"Why do rockets looks white?"	"Why are rockets and boosters painted white?"	Short-short

three benchmark collections for each application scenario and compare performance of models introduced in Sect. 3 respectively.

4.1 Datasets

For the Ad-hoc retrieval application, we choose a subset of Robust04. Robust04 is a dataset related to news and its topics are chosen from TREC Robust Track 2004. We use the description of each topic and a subset of these documents which has a headline. As for Web-QA application, we use Yahoo-QA dataset. The data is a subset of the Yahoo! Answers corpus from a 10/25/2007 dump. It is a small subset of the questions selected for their linguistic properties (for example they all start with "how $\{to|do|did|does|can|would|could|should\}$". For CQA application, we conduct our experiment on Quora dataset. Quora dataset consists of over 400,000 lines of potential question duplicate pairs. For all the datasets in our experiments, both queries and documents are stemmed using the Krovetz stemmer [13]. Stopwords is removed according to the INQUERY stop list. The statistics of these datasets are listed in Table 1.

Table 2. Statistics of the datasets used in this study.

	Yahoo-QA	Robust04	Quora
Vocabulary	53K	0.5M	77K
Document count	0.5M	0.34M	0.54M
Collection length	4.5M	474M	21M
Query count	9401	233	0.54M

4.2 Experimental Settings

Baseline Methods. We adopt BM25 as our baselines. We use the Indri to build index for documents and adopt the BM25TF achieved in the Indri as our baselines. We set the parameters as default in the Indri complemention, i.e. $k_1 = 1.2$ and $b = 0.75$ except for Robust04 datasets. Following the study of [12], we set $k_1 = 0.708$ and $b = 0.325$ for Robust04 dataset.

LM_DIR. The language model implemented in the Indri is based on a combination of the language modeling [22] and inference network [24] retrieval frameworks. Here we choose Dirichlet as smoothing method. We use default parameters, i.e., $\mu = 2500$ for Dirichlet.

Word Embeddings. To avoid randomness caused by out-of-vocabulary words, we adopt corpus-specific word embeddings in our experiment. We train word vectors using CBOW model [17] with negative sampling. We use Word2Vec[1] training 300-dimension word embeddings. We set the context window size to 10 and use 10 negative samples and the sampling threshold is 10^{-4} as default. The minimum count is set to 1 so that there won't be any word out-of-vocabulary.

Paragraph Vectors. The dimension of PV is 300. We use 6 negative samples and the sampling threshold is 10^{-4} as default and minimum count is set to 1.

Hybrid Models. We choose BM25 represents term based models, so there are totally four hybrid models namely $BM25 + BoWE_{score}$, $BM25 + BoWE_{rank}$, $BM25 + PV_{score}$ and $BM25 + PV_{rank}$ respectively. For coefficient used in each hybrid model, we choose the best from the range $(0, 1)$ increased by 0.05 each time according to experiments. Different hybrid model prefer different coefficient but the tendency are same. We show the coefficient tendency of the $BM25 + BoWE_{score}$ in Fig. 1.

Evaluation Measures. Since the initial retrieval step focus on retrieving as many related candidate documents as possible, thus we measure the model's performance in terms of recall@K only. We set different value of K for each datasets. Specifically, we set K to 500, 1000 and 100 for Yahoo-QA, Robust04 and Quora respectively.

---

[1] http://nlp.stanford.edu/projects/glove/.

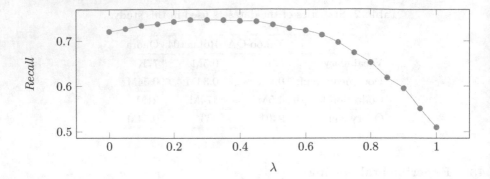

Fig. 1. Coefficient tendency of the $BM25 + BoWE_{score}$ Model

4.3   Retrieval Performance and Analysis

In this section we present our experimental results over three datasets. Note that we choose BM25TF as our baseline and "win", "tie", "lose" in each table mean that the number of cases win/tie/lose compared to baseline model.

Table 2 shows the result of models on Robust04 dataset which is corresponding to the application of Ad-hoc retrieval. In Robust04 dataset the queries is in short description form while the documents are long texts. The result shows that except for PV, all these models perform better than QA task. Since the relevant documents are required to describe the given topic, so the whole passage is taken into consideration. Although traditional models calculate relevance for the whole documents but due to the drawback of the BoW method, they suffer the vocabulary mismatch problem and loss semantic matching information, so the performance is not so well. Embedding based methods still can't do well in long texts but it do help in improving semantic relevance.

Table 3. Comparison of different initial retrieval methods over Robust04 for retrieving 1000 documents.

Robust04 collection (Recall@1000)					
Retrieval type	Retrieval method	Recall (%)	Win	Tie	Lose
Term based	BM25	72.07	–	–	–
	LM_DIR	72.52	54	126	53
Embedding based	BoWE	50.98	34	30	169
	PV	1.67	1	4	228
Hybrid	$BM25 + BoWE_{score}$	74.63	93	112	28
	$BM25 + PV_{score}$	72.67	55	166	12
	$BM25 + BoWE_{rank}$	74.41	78	129	26
	$BM25 + PV_{rank}$	71.69	5	186	42

Table 3 shows the result of different models on Yahoo-QA dataset which is corresponding to the application of QA. The problem of vocabulary mismatch and the requirement of finding the semantic relation between the queries and documents become more obviously in the QA scenario. Both queries and documents may be long texts in the QA tasks as in Yahoo-QA dataset. So the relevance measurement becomes more complicated. As the experimental result shows, the recall of each model stay in relative low level. Despite that, the performance of two models only based on embedding lag far behind traditional models, especially for PV. The reason may be that embedding based method aggregate whole text information together in a rough way which may lose many useful information for matching. Besides, since the length of queries is long, it calls for a method to understand the intension of the queries which is important in finding answers, but all these models fails in that. Further more, the answer for a given question is just a piece of words in document in most cases, so it is not proper to consider the relevance of whole documents. More Sophisticated models need to be used in this task.

Table 4 shows the result of different models on Quora dataset which is corresponding to the application of CQA. CQA tries to find the relevant question pairs, so the length of texts is relatively short. What's more, for Quora dataset, the problem of vocabulary mismatch in not so obviously, so even simply by traditional models based on BoW can gain pretty well performance. It can also be seen that methods based on embedding perform well too. When combined with traditional models, they contribute more (Table 5).

Table 4. Comparison of different initial retrieval methods over Yahoo-QA for retrieving 500 documents.

Yahoo-QA collection (Recall@500)					
Retrieval type	Retrieval method	Recall (%)	Win	Tie	Lose
Term based	BM25TF	55.45	–	–	–
	LM_DIR	53.98	156	8661	583
Embedding based	BoWE	33.25	859	3804	4737
	PV	8.84	264	1968	7168
Hybrid	$BM25 + BoWE_{score}$	56.61	825	8073	502
	$BM25 + PV_{score}$	55.02	231	8843	326
	$BM25 + BoWE_{rank}$	56.82	900	7953	547
	$BM25 + PV_{rank}$	55.52	345	8745	310

Table 5. Comparison of different initial retrieval methods over Quora for retrieving 100 documents.

Quora collection (Recall@100)					
Retrieval type	Retrieval method	Recall (%)	Win	Tie	Lose
Term based	BM25TF	94.27	–	–	–
	LM_DIR	87.66	596	134290	14708
Embedding based	BoWE	89.69	6279	128535	14780
	PV	84.96	4615	121112	23867
Hybrid	$BM25 + BoWE_{score}$	95.42	5112	143408	1074
	$BM25 + PV_{score}$	94.66	4422	142976	2196
	$BM25 + BoWE_{rank}$	95.49	5453	142351	1790
	$BM25 + PV_{rank}$	94.97	4153	143548	1893

Now let's overview the all experimental results to draw some general conclusions.

First of all, comparing the performance of various traditional methods, we can see that in the initial retrieval step, different traditional methods perform nearly equally. But BM25 outperform language model with different smoothing methods slightly in general. The performance of these models are equal on most cases.

Next, when it comes to the methods based on embeddings, we can see that both two methods can't outperform the traditional methods. PV perform extremely poor in long texts scenario. Similarly, BoWE also perform badly in long texts scenario. Long texts contain too many information to be considered so it is hard to aggregate all these information in a simple way like PV or BoWE. Although embedding based method perform pretty well in many NLP tasks, in IR tasks there need some enhancement to fit the special requirements.

Finally, we can find that embedding based method do better in semantic matching. The retrieval results of embedding based methods are quite different from traditional methods. In other words, their advantages are not overlapped. So when combine the traditional models with embedding based models simply through linear combination, the performance still can be improved. Through the combination, we add some semantic relevance information into merely words co-occurrence (i.e., exact matching). What's more, the combination by ranking position is better than combination through score in most cases.

5   Conclusion

In this paper, we conduct experiments on different IR application tasks and draw the conclusion that embedding based models do help in improving the recall of traditional initial retrieval models. We confirm that embedding based methods BoWE representation and Paragraph Vector can help traditional model in the cases that queries and documents are related in semantic level which share little common words. Through two combination methods, we illustrate that traditional retrieval models combined with embedding based models can perform better than the single one.

For future work, we would like to find a better method to capture semantic relevance information and further more, come up with more effective method to combine the advantage of exact matching models and semantic matching models.

References

1. Blooma, M.J., Kurian, J.C.: Research issues in community based question answering. In: PACIS, p. 29 (2011)
2. Burges, C., et al.: Learning to rank using gradient descent. In: Proceedings of the 22nd International Conference on Machine Learning, pp. 89–96. ACM (2005)
3. Cao, Z., Qin, T., Liu, T.-Y., Tsai, M.-F., Li, H.: Learning to rank: from pairwise approach to listwise approach. In: Proceedings of the 24th International Conference on Machine Learning, pp. 129–136. ACM (2007)
4. Carmel, D., Mejer, A., Pinter, Y., Szpektor, I.: Improving term weighting for community question answering search using syntactic analysis. In: Proceedings of the 23rd ACM International Conference on Information and Knowledge Management, pp. 351–360. ACM (2014)
5. Dang, V., Bendersky, M., Croft, W.B.: Two-stage learning to rank for information retrieval. In: Serdyukov, P. (ed.) ECIR 2013. LNCS, vol. 7814, pp. 423–434. Springer, Heidelberg (2013). https://doi.org/10.1007/978-3-642-36973-5_36
6. Firth, J.R.: A synopsis of linguistic theory, 1930–1955. In: Studies in Linguistic Analysis (1957)
7. Galitsky, B.: Natural language question answering system: technique of semantic headers. Advanced Knowledge International (2003)
8. Guo, J., Fan, Y., Ai, Q., Croft, W.B.: A deep relevance matching model for ad-hoc retrieval. In: Proceedings of the 25th ACM International on Conference on Information and Knowledge Management, pp. 55–64. ACM (2016)
9. Guo, J., Fan, Y., Ai, Q., Croft, W.B.: Semantic matching by non-linear word transportation for information retrieval. In: Proceedings of the 25th ACM International on Conference on Information and Knowledge Management, pp. 701–710. ACM (2016)
10. Harris, Z.S.: Distributional structure. Word 10(2–3), 146–162 (1954)
11. Huang, P.-S., He, X., Gao, J., Deng, L., Acero, A., Heck, L.: Learning deep structured semantic models for web search using clickthrough data. In: Proceedings of the 22nd ACM International Conference on Information and Knowledge Management, pp. 2333–2338. ACM (2013)
12. Huston, S., Croft, W.B.: Parameters learned in the comparison of retrieval models using term dependencies. Ir, University of Massachusetts (2014)
13. Krovetz, R.: Viewing morphology as an inference process. In: Proceedings of the 16th Annual International ACM SIGIR Conference on Research and development in Information Retrieval, pp. 191–202. ACM (1993)
14. Le, Q., Mikolov, T.: Distributed representations of sentences and documents. In: International Conference on Machine Learning, pp. 1188–1196 (2014)

15. Lin, J.: The web as a resource for question answering: perspectives and challenges. In: Proceedings of the Third International Conference on Language Resources and Evaluation (LREC 2002). Citeseer (2002)

16. Lu, Z., Li, H.: A deep architecture for matching short texts. In: Advances in Neural Information Processing Systems, pp. 1367–1375 (2013)

17. Mikolov, T., Sutskever, I., Chen, K., Corrado, G.S., Dean, J.: Distributed representations of words and phrases and their compositionality. In: Advances in Neural Information Processing Systems, pp. 3111–3119 (2013)

18. Pang, L., Lan, Y., Guo, J., Xu, J., Xu, J., Cheng, X.: DeepRank: a new deep architecture for relevance ranking in information retrieval. In: Proceedings of the 2017 ACM on Conference on Information and Knowledge Management, pp. 257–266. ACM (2017)

19. Patra, B.: A survey of community question answering. arXiv preprint arXiv:1705.04009 (2017)

20. Pennington, J., Socher, R., Manning, C.: GloVe: global vectors for word representation. In: Proceedings of the 2014 Conference on Empirical Methods in Natural Language Processing (EMNLP), pp. 1532–1543 (2014)

21. Perera, R.: IPedagogy: question answering system based on web information clustering. In: 2012 IEEE Fourth International Conference on Technology for Education (T4E), pp. 245–246. IEEE (2012)

22. Ponte, J.M., Croft, W.B.: A language modeling approach to information retrieval. In: Proceedings of the 21st Annual International ACM SIGIR Conference on Research and Development in Information Retrieval, pp. 275–281. ACM (1998)

23. Robertson, S.E., Walker, S.: Some simple effective approximations to the 2-poisson model for probabilistic weighted retrieval. In: Croft, B.W., van Rijsbergen, C.J. (eds.) SIGIR '94, pp. 232–241. Springer, New York (1994). https://doi.org/10.1007/978-1-4471-2099-5_24

24. Turtle, H., Croft, W.B.: Evaluation of an inference network-based retrieval model. ACM Trans. Inf. Syst. (TOIS) 9(3), 187–222 (1991)

25. Vulić, I., Moens, M.-F.: Monolingual and cross-lingual information retrieval models based on (bilingual) word embeddings. In: Proceedings of the 38th International ACM SIGIR Conference on Research and Development in Information Retrieval, pp. 363–372. ACM (2015)

26. Zhai, C., Lafferty, J.: Two-stage language models for information retrieval. In: Proceedings of the 25th Annual International ACM SIGIR Conference on Research and Development in Information Retrieval, pp. 49–56. ACM (2002)

27. Zhai, C., Lafferty, J.: A study of smoothing methods for language models applied to ad hoc information retrieval. In: ACM SIGIR Forum, vol. 51, pp. 268–276. ACM (2017)

Text Matching with Monte Carlo Tree Search

Yixuan He, Shuchang Tao, Jun Xu$^{(\boxtimes)}$, Jiafeng Guo, YanYan Lan,
and Xueqi Cheng

CAS Key Lab of Network Data Science and Technology,
Institute of Computing Technology, Chinese Academy of Sciences,
Beijing, China
{heyixuan,taoshuchang}@software.ict.ac.cn,
{junxu,guojiafeng,lanyanyan,cxq}@ict.ac.cn

Abstract. In this paper we address a novel reinforcement learning based model for text matching, referred to as MM-Match. Inspired by the success and methodology of the AlphaGo Zero, MM-Match formalizes the problem of text matching with a Monte Carlo tree search (MCTS) enhanced Markov decision process (MDP) model, where the time steps corresponded to the positions in a match matrix from top left corner to down right corner, and each action corresponds to a movement in a direction. Two long short-term memory networks (LSTM) are used to summarize the words in current path and one more LSTM is used to summarize future words. Based on the outputs of LSTMs, the policy guides the move direction and the value predicts the correctness of the whole sentence are produced. The policy and value are then strengthened with MCTS, which takes the produced raw policy and value as inputs, simulates and evaluates the possible direction assignments at the subsequent positions, and outputs a better search policy for assigning directions. A reinforcement learning algorithm is proposed to train the model parameters. Our work innovatively applies an MDP model to the text matching task. MM-Match can accurately predict the directions thanks to the exploratory decision making mechanism introduced by MCTS. Experimental results showed that MM-Match performs similar to the classical text matching models including MatchPyramid and MatchSRNN.

Keywords: Monte Carlo tree search · Markov decision process
Text matching

1 Introduction

Text matching is a basic task in natural language understanding, which can be applied to many natural language processing tasks, including question answering, information retrieval and paraphrase identification. Taking paraphrase identification as an example, given two sentences (e.g. **Q** and **D**), the matching model determine whether **Q** and **D** match or not.

© Springer Nature Switzerland AG 2018
S. Zhang et al. (Eds.): CCIR 2018, LNCS 11168, pp. 41–52, 2018.
https://doi.org/10.1007/978-3-030-01012-6_4

In this paper we propose to solve the text matching with a Monte Carlo tree search (MCTS) enhanced Markov decision process (MDP) [10], inspired by the reinforcement learning model of AlphaGO [8] and AlphaGO Zero [9] programs designed for the Game of Go, The new text matching model, referred as MM-Matching (MCTS enhanced MDP for Matching), uses an MDP to model the path generating process in text matching. We still need a match matrix, but each position in matrix stands for a states in MDP. We start from the top left corner and walk through the matrix until reach the down right corner. At each time, based on the past path and k forward words, two LSTMs are used to summarize the path and an LSTM is used to summarize forward words, respectively. Based on the outputs of these three LSTMs, there is a policy function (a distribution over the valid direction) for guiding direction for next movement and a value function for estimating the correctness of label we predict. To avoid the problem of moving without utilizing the whole sentence information, MM-Matching explores more possibilities in the whole space instead of choosing a direction directly with the raw policy predicted by the policy function. The exploration is conducted with the MCTS guided by the produced policy function and value function, resulting a strengthened search policy for the text matching. Moving to the next iteration, the algorithm moves to the next position and continues the above process until reaches the down right corner.

Reinforcement learning is used to train the model parameters. In the training phase, at each learning iteration and for each training sentence pair (and the corresponding labels), the algorithm first conducts an MCTS inside the training loop, guided by the current policy function and value function. Then the model parameters are adjusted to minimize the loss function. The loss function consisted of two terms: (1) the cross entropy between the predicted value and the correctness of model output; (2) the cross entropy of the predicted policy and the search probabilities for directions selection. Stochastic gradient descent is utilized for conducting the optimization.

To evaluate the effectiveness of MM-Matching, we conducted experiments on the basis of Quora question dataset. The experimental results showed that MM-Matching can significantly outperform the state-of-the-art text matching approaches, including the MatchPyramid [5] and MatchSRNN [11]. We analyzed the results and showed that MM-Matching improved the performances through conducting lookahead MCTS to explore in the whole matching space.

2   Related Work

Text matching models can mainly be divided into traditional probabilistic models and deep neural networks models. Traditional models [1] need a large number of hand-crafted features but few parameters that can be learned, which limits the generalization of the models. The deep learning method can automatically extract features from the original data eliminating labor overhead. At the same time, the deep text matching model combined with the Word2Vec solves multiple level problem of text matching. Deep Semantic Structured Model (DSSM) [3] and

Convolutional Deep Semantic Structured Model (CDSSM) [7] generate word vectors by neural network and calculates similarity between the two vectors which means the match degree. It still difficult to capture the context information that is far apart. LSTM-DSSM [4] uses LSTM to solve the above problem. However, these DSSM models need a large number of training datas and its performance is quite uncontrollable. MatchPyramid [5] presents a match matrix to model the interactions of sentence pairs. It can capture important matching patterns such as unigram, n-gram and n-term in different levels. MatchSRNN [11] models the recursive matching structure to calculate the matching degree. It also find a matching path between two sentences by backtracking. Inspired by this, we intend to find the matching path because it reflects the inside match pattern and is more in line with human judgment on whether two sentences match. However, it needs a lot of features and labeled data to define how to walk through the match matrix correctly if tackled by supervise learning such as using traditional and deep learning methods.

Recently, reinforcement learning has achieved a great success in the game area as it performs well in sequential decision making problem. Reinforcement learning models automatically interact with the environment and collecting data while running. Therefore, it naturally meets the need of finding a matching path.

3   MDP Formulation of Text Matching

This section introduces our text matching model. Inspired by MatchSRNN, we want to find a path between two sentences and determine whether they match or not based on this path. But it is a huge challenge to generate a path without any human knowledge, so we use reinforcement learning. Our algorithm pioneers the use of reinforcement learning to find a path, and then output a match result based on this path (Fig. 1).

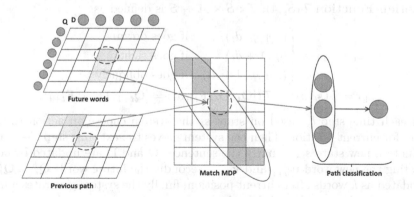

Fig. 1. The architecture of MM-Match

3.1   Text Matching as an MDP

Given two sentence $\mathbf{Q} = \{\mathbf{u}_1, \mathbf{u}_2, \cdots, \mathbf{u}_n\}, \mathbf{D} = \{\mathbf{v}_1, \mathbf{v}_2, \cdots, \mathbf{v}_m\}$, the target of text matching is to learn a match model $f : \mathbf{Q} \times \mathbf{D} \to \mathbb{R}$, which can be applied for any sentence $q \in \mathbf{Q}, d \in \mathbf{D}$ for computing similarity. All component of w_i and v_i are the L-dimensional representation of words, i.e. the word embedding.

Composition of MDP. Pattern text matching formulates the matching of two sentence as a process of sequential decision making with a MDP where each step corresponding to a position in a match matrix. The model, is represented by a tuple $<\mathcal{S}, \mathcal{A}, \mathcal{T}, V, \pi>$ composed by states, actions, transition, value, and policy, which are respectively defined as follows: The states, actions, transition function, value function, and policy function of the MDP are set as:

States $\mathcal{S}$: We design the state at time step t as the path we already walked through the match matrix.

$$\mathbf{Q}_t = \{\mathbf{u}_{q_1}, \mathbf{u}_{q_2}, \mathbf{u}_{q_3}, \cdots, \mathbf{u}_{q_t}\},$$
$$\mathbf{D}_t = \{\mathbf{v}_{d_1}, \mathbf{v}_{d_2}, \mathbf{v}_{d_3}, \cdots, \mathbf{v}_{d_t}\},$$
$$\mathbf{F}_t = \{[\mathbf{u}_{q_t}, \mathbf{v}_{d_t}], [\mathbf{u}_{1+q_t}, \mathbf{v}_{1+d_t}] \cdots, [\mathbf{u}_{k+q_t}, \mathbf{v}_{k+d_t}]\},$$
$$s_t = [\mathbf{Q}_t, \mathbf{D}_t, \mathbf{F}_t]$$

where q_t and d_t is the current position in the match matrix, $\mathbf{Q}_t$ and $\mathbf{D}_t$ is the preliminary representation of the prefix of the input text of length q_t and d_t. $\mathbf{F}_t$ consists of the future words pairs. At the beginning ($t = 1$), the state is initialized with position $[1, 1]$, as $s_1 = [\mathbf{Q}_1 = \{\mathbf{u}_1\}, \mathbf{D}_1 = \{\mathbf{v}_1\}, \mathbf{F}_1 = \{[\mathbf{u}_1, \mathbf{v}_1], \cdots, [\mathbf{u}_k, \mathbf{v}_k]\}]$.

Actions $\mathcal{A}$: At each time step t, we can choose to go right, down or obliquely. Each time we choose a direction to go and change position according to that direction.

Transition Function $\mathcal{T}(S, \mathcal{A})$: $\mathcal{T} : S \times \mathcal{A} \to S$ is defined as

$$(q_{t+1}, d_{t+1}) = \begin{cases} (1 + q_t, d_t) & \text{if goes down} \\ (q_t, 1 + d_t) & \text{if goes right} \\ (1 + q_t, 1 + d_t) & \text{if goes obliquely} \end{cases}$$
$$s_{t+1} = \mathcal{T}(s_t, a_t) = \mathcal{T}([\mathbf{Q}_t, \mathbf{D}_t, \mathbf{F}_t], a_t) = [\mathbf{Q}_{t+1}, \mathbf{D}_{t+1}, \mathbf{F}_{t+1}]$$

At each time step t, based on state s_t the system choose an action (direction) a_t for current position. Then the system moves to next time step $t + 1$ and transits to a new state s_{t+1}: firstly the sentences $\mathbf{Q}$ and $\mathbf{D}$ are updated by concatenating the new word $\mathbf{u}_{q_{t+1}}$ and $\mathbf{v}_{d_{t+1}}$, secondly the future words $\mathbf{Q}_{t+1}^F, \mathbf{Q}_{t+1}^F$ are updated as k words after current position, finally the system generates a new direction according to selected direction,

Value Function V: The state value function $V : S \to \mathbb{R}$ is a scalar evaluation, predicting whether this model can obtain right result for two sentences. The value function is learned so as to fit the real result of the training sentences.

In this paper, two LSTMs are used to map the word sequence $\mathbf{Q}_t, \mathbf{D}_t$ to two vectors respectively. We define the value functions as nonlinear transformation of the weighted sums of the LSTM's output:

$$V(s) = \sigma(<\mathbf{w}, \mathbf{g}_v(s)> + \mathbf{b}_v) \qquad (1)$$

where $\mathbf{w}, \mathbf{b}_v$ are the weight vector and the bias to be learned during training. $\sigma(x) = \frac{1}{1+e^{-x}}$ is the nonlinear sigmoid function, and $\mathbf{g}_v(s)$ is a concatenation of the outputs from the sentence LSTM $\text{LSTM}_Q, \text{LSTM}_D$:

$$\mathbf{g}_v(s) = [\text{LSTM}_Q(\mathbf{Q}_t)^T, \text{LSTM}_D(\mathbf{D}_t)^T]^T \qquad (2)$$

Those LSTM networks are defined as followings:

$$s = [\mathbf{Q}_t = \{\mathbf{u}_{q_1}, \cdots, \mathbf{u}_{q_t}\}, \mathbf{D}_t = \{\mathbf{v}_{d_1}, \cdots, \mathbf{v}_{d_t}\}$$

where $u_{q_k} (k = 1, \cdots, t)$ and $v_{q_k} (k = 1, \cdots, t)$ is the word at k-th position, represented with its word embedding. LSTM_Q outputs a representation h_k for position k. The output vector and cell state vector at the t-th cell are concatenated as the output of LSTM, that is

$$\text{LSTM}_Q(\mathbf{Q}_t) = \left[\mathbf{h}_t{}^T, \mathbf{c}_t{}^T\right]^T.$$

The function $\text{LSTM}_D(\mathbf{D}_t)$, which is used to map another sentence $\mathbf{D}_t$ into a real vector, are defined similarly to that for LSTM_Q. LSTM_D and LSTM_Q are sharing weights.

Policy Function π: $\pi(s)$ defines a function that takes the state as an input and outputs a distribution over all of the possible actions $a \in \mathcal{A}(s)$. The policy function takes the future vector $\mathbf{F}_t$ and bilinear product of the state representation in Eq. (1) as input. The function uses one LSTM to map $\mathbf{F}_t$ into a vector and concatenates it with $\mathbf{g}_v(s)$. Each probability in the distribution is a normalized soft-max function of the LSTM's output:

$$\pi(a|s) = \frac{\exp\left\{\Phi(a)^T\mathbf{U}_\pi \, \mathbf{g}_\pi(s)\right\}}{\sum_{a'\in\mathcal{A}(s)} \exp\left\{\Phi(a')^T\mathbf{U}_\pi \, \mathbf{g}_\pi(s)\right\}} \qquad (3)$$
$$\mathbf{g}_\pi(s) = [\mathbf{g}_v(s), \text{LSTM}_F(\mathbf{F}_t)^T]^T$$

where $\Phi(a)$ is a one hot vector for representing each direction a and $\mathbf{U}_p$ is the parameter in bilinear product. The policy function $\pi(s)$ is:

$$\pi(s) = \langle\pi(a_1|s), \cdots, \pi(a_{|\mathcal{A}(s)|}|s)\rangle. \qquad (4)$$

The LSTM_F is similarly to LSTM_Q

Path Classification Model. When a path is found, we need a model to classify whether this path is a correct matching path. This model takes $\mathbf{Q}_t$ and $\mathbf{D}_t$ as input and output a probability whether the two sentences match. In this paper, there are two LSTMs for mapping two sequence $\mathbf{Q}_t$ and $\mathbf{D}_t$ into two matrix, and one LSTM taking these two matrix as input.

The output probability is normalized sigmoid function of the LSTM's output:

$$p_m = \frac{1}{1 + e^{-\mathbf{g}_c^{\text{out}}}} \tag{5}$$

LSTM_Q and LSTM_D share weights.

3.2   Strengthening Raw Policy with MCTS

Matching directly with policy π in Eq. (4) or value $\mathbf{V}$ in Eq. (1) is easily lead to suboptimal because the policy and value is computed based on history path and a short future words. The raw policy doesn't contain any information about the movement afterwards and the whole sentences. To solve this problem, following the practices in AlphaGo [8] and AlphaGo Zero [9], we use MCTS to conduct lookahead search. That is, we will execute an MCTS search at time t with policy function π and value function $\mathbf{V}$, and output a strengthened new search policy π. Usually, the search policy π has higher probability to select a direction with high accuracy than the raw policy π defined in Eq. (4).

Algorithm 1 shows the details of the MCTS in which each tree node corresponds to an MDP state. It takes a root node s_R, value function $\mathcal{V}$ and policy function π as inputs. The algorithm iterates K times and outputs a strengthened search policy $\boldsymbol{\pi}$ for selecting a direction for the root node s_R. Suppose that each edge $e(s, a)$ (the edge from state s to the state $\mathcal{T}(s, a)$) of the MCTS tree stores an action value $Q(s, a)$, visit count $N(s, a)$, and prior probability $\pi(s, a)$. At each of the iteration, the MCTS executes the following steps:

Selection: Each iterations starts from the root state s_R and iteratively selects the documents that maximize an upper confidence bound:

$$a_t = \arg\max_a (Q(s_t, a) + \lambda U(s_t, a)) \tag{6}$$

where $\lambda > 0$ is trade-off coefficient, and bonus $U(s_t, a)$:

$$U(s_t, a) = p(a|s_t) \frac{\sqrt{\sum_{a' \in \mathcal{A}(s_t)} N(s_t, a')}}{1 + N(s_t, a)}$$

$U(s_t, a)$ is proportional to the prior probability but decays with repeated visits to encourage exploration.

Evaluation and Expansion: When the traversal reaches a leaf node s_L, the node is evaluated with the value function $V(s_L)$ (Eq. (1)). Note following the practices in AlphaGo Zero, we use the value function for evaluating a node.

Then, the leaf node s_L may be expanded. Each edge from the leaf position s_L (corresponds to each action $a \in \mathcal{A}(s_L)$) is initialized as: $\pi(s_L, a) = \pi(a|s_L)$ (Eq. (4)), $Q(s_L, a) = 0$, and $N(s_L, a) = 0$. In this paper all of the available actions of s_L are expanded.

Algorithm 1. TreeSearch

Input: root s_R, value V, policy π, search times K
Output: Search policy π
 1: **for** $k = 0$ to $K - 1$ **do**
 2:    $s_L \leftarrow s_R$
 3:    {Selection}
 4:    **while** s_L is not a leaf node **do**
 5:       $a \leftarrow \arg\max_{a \in \mathcal{A}(s_L)} Q(s_L, a) + \lambda \cdot U(s_L, a)$ {Eq. (6)}
 6:       $s_L \leftarrow$ child node pointed by edge (s_L, a)
 7:    **end while**
 8:    {Evaluation and expansion}
 9:    $v \leftarrow V(s_L)$ {simulate v with value function V}
10:    **for all** $a \in \mathcal{A}(s_L)$ **do**
11:       Expand an edge e to node $s = [s_L.\mathbf{Q}_t, s_L.\mathbf{D}_t, s_L.\mathbf{F}_t]$
12:       $e.P \leftarrow p(a|s_L); e.Q \leftarrow 0; e.N \leftarrow 0$ {init edge properties}
13:    **end for**
14:    {Back-propagation}
15:    **while** $s_L \neq s_R$ **do**
16:       $s \leftarrow$ parent of $s_L; e \leftarrow$ edge from s to s_L
17:       $e.Q \leftarrow \frac{e.Q \times e.N + v}{e.N + 1}$ {Equation (7)}
18:       $e.N \leftarrow e.N + 1; s_L \leftarrow s$
19:    **end while**
20: **end for**
21: {Calculate tree search policy. Eq. (8)}
22: **for all** $a \in \mathcal{A}(s_R)$ **do**
23:    $\pi(a|s_R) \leftarrow \frac{e(s_R, a).N}{\sum_{a' \in \mathcal{A}(s_R)} e(s_R, a').N}$
24: **end for**
25: **return** π

Back-Propagation and Update: At the end of evaluation, the action values and visit counts of all traversed edges are updated. For each edge $e(s, a)$, the prior probability $\pi(s, a)$ is kept unchanged, and $Q(s, a)$ and $N(s, a)$ are updated:

$$Q(s, a) \leftarrow \frac{Q(s, a) \times N(s, a) + V(s_L)}{N(s, a) + 1}; N(s, a) \leftarrow N(s, a) + 1. \qquad (7)$$

Calculate the Strengthened Search Policy: Finally after iterating K times, the strengthened search policy π for the root node s_R can be calculated according to the visit counts $N(s_R, a)$ of the edges starting from s_R:

$$\pi(a|s_R) = \frac{N(s_R, a)^{\frac{1}{\tau}}}{\sum_{a' \in \mathcal{A}(s_R)} N(s_R, a')^{\frac{1}{\tau}}}, \qquad (8)$$

for all $a \in \mathcal{A}(s_R)$, where $\tau > 0$ is system temperature.

3.3   Learning and Inference Algorithms

Reinforcement Learning of the Parameters. The model has parameters Θ_{MCTS} (including $\mathbf{w}, b_v, \mathbf{U}_p$, and parameters in $LSTM_D$, $LSTM_Q$ and $LSTM_F$), Θ_c(including parameters in $LSTM_D$, $LSTM_Q$ and $LSTM_c$) to learn. In the training phase, suppose we are given N labeled sentence pair $D = \{(\mathbf{Q}^{(n)}, \mathbf{D}^{(n)}), \mathbf{Y}\}_{n=1}^{N}$. Algorithm 2 shows the training procedure. First, the parameters Θ_{MCTS} and Θ_c are initialized to random weights in $[-1, 1]$. At each subsequent iteration, for each sentence pair $(\mathbf{Q}, \mathbf{D})$, a path is predicted for $\mathbf{Q}$ and $\mathbf{D}$ with current parameter setting: at each position t, an MCTS search is executed, using previous iteration of value function and policy function, and a direction a_t is selected according to the search policy $\boldsymbol{\pi}_t$. The matching terminates when reach the down right corner of the match matrix and achieves a predicted path sequence $([u_{q_1}, u_{q_2}, \cdots, u_{q_t}], [v_{d_1}, v_{d_2}, \cdots, v_{d_t}])$. Classification model is updated with path sequence and ground truth Y with loss:

$$\ell_c(Y, p) = -\left(Y \log(p) + (1 - Y) \log(1 - p)\right) \tag{9}$$

After updating classification model, prediction result Y_p is calculated with each path. Then reward $r = \mathbb{I}_{Y=Y_p}$ is computed. The data generated at each time step $E = \{(s_t, \boldsymbol{\pi}_t)\}_{t=1}^{M}$ and the final evaluation r are utilized as the signals in training for adjusting the value function. The model parameters are adjusted to minimize the error between the predicted value $V(s_t)$ and predict result r, and to maximize the similarity of the policy $\mathbf{p}(s_t)$ to the search probabilities $\boldsymbol{\pi}_t$. Specifically, the parameters Θ_{MCTS} are adjusted by gradient descent on a loss function ℓ that sums over the mean-squared error and cross-entropy losses, respectively:

$$
\ell(E, r) = \sum_{t=1}^{|E|} \left(-(r \log V(s_t) + (1 - r) \log(1 - V(s_t))) \right) \\
+ \beta \sum_{a \in \mathcal{A}(s_t)} \pi_t(a|s_t) \log \frac{1}{\pi(a|s_t)}. \tag{10}
$$

The model parameters are trained by back propagation and stochastic gradient descent. Specifically, we use AdaGrad [2] on all parameters in the training process.

Inference. The inference of the text matching for a sentence pair is shown in Algorithm 3. Given a sentence pair $\mathbf{Q}, \mathbf{D}$, the system state is initialized as $s_1 = [\{\mathbf{u}_1\}, \{\mathbf{v}_1\}, \{[\mathbf{u}_1, \mathbf{v}_1], \cdots, [\mathbf{u}_k, \mathbf{v}_k]\}$. Then, at each of the time steps $t = 1, \cdots, M$, the agent receives the state $s_t = [\mathbf{Q}_t, \mathbf{D}_t, \mathbf{F}_t]$ and search the policy $\boldsymbol{\pi}$ with MCTS, on the basis of the value function V and policy function $\mathbf{p}$. Then, it chooses an action a for the word at position t. Moving to the next iteration $t + 1$, the state becomes $s_{t+1} = [\mathbf{Q}_{t+1}, \mathbf{D}_{t+1}, \mathbf{F}_{t+1}]$. The process is repeated until the down right corner of the matching matrix.

We implemented the Text matching model based on TensorFlow and the code can be found at the Github repository http://hide_for_anonymous_review.

Algorithm 2. Train text matching model

Input: Labeled data $D = \{(\mathbf{D}^{(n)}, \mathbf{Q}^{(n)}, \mathbf{Y}^{(n)})\}_{n=1}^{N}$, learning rate η, number of search
    K

Output: Θ_{MCTS}, Θ_c

 1: Initialize $\Theta_{MCTS}, \Theta_c \leftarrow$ random values in $[-1, 1]$
 2: **repeat**
 3:    **for all** $(\mathbf{Q}, \mathbf{D}, \mathbf{Y}) \in D$ **do**
 4:       $s \leftarrow [\{\mathbf{u}_1\}, \{\mathbf{v}_1\}, \{[\mathbf{u}_1, \mathbf{v}_1], \cdots, [\mathbf{u}_k, \mathbf{v}_k]\}]$
 5:       **while** not reach right down corner **do**
 6:          $\pi \leftarrow$ TreeSearch$(s, V, \mathbf{p}, K)$ {Alg. (1)}
 7:          sample action $a \in \mathcal{A}(s)$ according to $\pi(a|s)$
 8:          $E \leftarrow E \oplus \{(s, \pi)\}$
 9:          $s \leftarrow [s.\mathbf{Q}_{t+1}, s.\mathbf{D}_{t+1}, s.\mathbf{F}_{t+1}]$
10:       **end while**
11:       **while** not convergence **do**
12:          $\Theta_c \leftarrow \Theta_c - \eta \frac{\partial \ell_c(\mathbf{Y}, p)}{\partial \Theta_c}$ {ℓ_c is defined in Eq. (9)}
13:       **end while**
14:       compute $\mathbf{Y}_p$ by Eq. (5)
15:       $r \leftarrow \mathbb{I}_{\mathbf{Y} = \mathbf{Y}_p}$
16:       $\Theta_{MCTS} \leftarrow \Theta_{MCTS} - \eta \frac{\partial \ell(E, r)}{\partial \Theta_{MCTS}}$ {ℓ is defined in Eq. (10)}
17:    **end for**
18: **until** converge
19: **return** Θ_{MCTS}, Θ_c

---

4 Experiments

4.1 Datasets

In this section, we conduct the experiment on paraphrase identification and test the performance of MM-Match on the Quora dataset[1]. The dataset contains more than $400\,\mathrm{K}$ pairs of questions. Each pair is labeled whether they duplicated or not. We use 5-fold Cross Validation to avoid overfitting and underfitting. 240K for training 80 K for validation 80 K for testing. Considering MM-Match is time consuming for processing the long sentence, we constructed a short sentence subset which was randomly selected from QuoraQP dataset and the sentences between 3 than 50 words were kept. The final sentence pair subset consist of 12000 sentence pairs, 10000 of them is used for training and 2000 were used for testing. All of the words in the sentences were represented with the word embeddings. In the experiments, we used the publicly available GloVe 300-dimensional embeddings trained on 6 billion words from Wikipedia and Gigaword [6].

We compare MM-Match with MatchPyramid and MatchSRNN implemented with an open software MatchZoo.

In our experiments, we set the parameters for MM-Match as follows. The number of search times K is set to 200, the learning rate η is 0.001, β and λ is different in each fold.

---

[1] https://www.kaggle.com/c/quora-question-pairs.

Algorithm 3. Text matching Inference

Input: sentence pair $\mathbf{Q} = \{\mathbf{u}_1, \cdots, \mathbf{u}_M\}$, $\mathbf{D} = \{\mathbf{w}_1, \cdots, \mathbf{w}_M\}$, value function V, policy
   function $\mathbf{p}$, and search times K,
Output: label $\mathbf{Y}$
1: $s \leftarrow [\{\mathbf{u}_1\}, \{\mathbf{v}_1\}, \{[\mathbf{u}_1, \mathbf{v}_1], \cdots, [\mathbf{u}_k, \mathbf{v}_k]\}]$
2: **while** not reach right down corner **do**
3:   $\pi \leftarrow \text{TreeSearch}(s, V, \mathbf{p}, K)$
4:   $a \leftarrow \arg\max_{a \in \mathcal{A}(s)} \pi(a|s)$
5:   $s \leftarrow [s.\mathbf{Q}_{t+1}, s.\mathbf{D}_{t+1}, s.\mathbf{F}_{t+1}]$
6: **end while**
7: compute $\mathbf{Y}_p$ by Eq. (5)
8: **return** $\mathbf{Y_p}$

Table 1. Performance comparison of all methods.

	Accuracy	F1	Auc
MM-Match	71.92%	73.05%	78.23%
MatchSRNN	**72.37(*)%**	**74.40%(*)**	**79.34%(*)**
MatchPyramid	71.18%	72.27%	77.45%

The text matching can be formulated as classification task. We use typical classification measures, such as Accuracy, F1 and Area under curve of ROC (denoted as Auc). The experimental results are listed in Table 1. Boldface indicates the highest scores among all runs. We can see that although MatchSRNN performs best, MM-Match can improve the performances, compared with MatchPyramid. MM-Match also convergence faster than the other two methods. MM-Match takes about 4 h to run one epoch. The previous method such as MatchSRNN have to take the whole matrix into consideration and its time complexity is n^2. Our method only need to consider the matching path and reduce the complexity to n. In the future work, we will improve the algorithm efficiency, and try to apply MDP to matching problem in other approaches and further improve our models.

We conduct experiments to see how the number of lookahead word influence the performance of MM-Match. The result are listed in Table 2. Since it takes a long time to run the algorithm, we use part of the QuoraQP dataset without 5-fold cross validation to make it faster. Therefore, the Auc of MM-Match in Table 1 is less than that in Table 2.

Table 2. The influence of lookahead words count to our algorithm

	Accuracy	F1	Auc
$k = 2$	71.27%	**73.13%(*)**	72.92%
$k = 3$	**71.92%(*)**	73.05%	**74.40%(*)**
$k = 4$	71.57%	72.28	72.99%
$k = 5$	71.26%	72.66	72.86%

Boldface indicates the highest scores among all runs. We can see that when lookahead 3 words, MM-Match get best performance. It is because the number of words that cause the combination structure of languages problems is limited, and they are generally around 3 or 4. When $k = 2$, most of the orderly reversed scenes cannot be covered. When k is greater than 3, the algorithm performance become worse. This may be because when k is too large, the length of lookahead words is much larger than the length of the reversed order, so that the information captured by the neural network contains more sequence information, and the proportion of reversal of the information decreased.

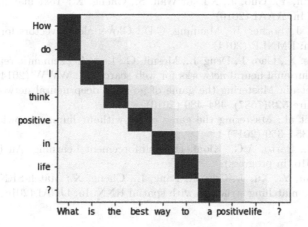

Fig. 2. Visit times of each node in Monte Carlo tree search

We visualize the visit times of each tree search node in MM-Match in order to analyze visit situation. Figure 2 shows each node visit times after the first tree search with the input sentence pair **How do I think positive in In the case of life?"** and **What is the best way to a positive life?**. We can see that the upper right part and the lower left part are obviously empty, and the number of visits at the diagonal is the highest. It meets the natural behavior of human.

5   Conclusion

In this paper we have proposed a novel approach to text matching, referred to as MM-Match. MM-Match formalizes the problem of text matching with a Monte Carlo tree search (MCTS) enhanced Markov decision process (MDP). The lookahead MCTS is used to strengthen the raw predicted policy so that the search policy has high probability to select the correct for each word. Reinforcement learning is utilized to train the model parameters. MM-Match enjoys several advantages: matching with the shared policy and the value functions, end-to-end learning, and high accuracy in matching. Experimental results show that MM-Match performs similar to the baselines of MatchSRNN and Match-Pyramid.

References

1. Das, D., Smith, N.A.: Paraphrase identification as probabilistic quasi-synchronous recognition. In: ACL/IJCNLP (2009)
2. Duchi, J.C., Hazan, E., Singer, Y.: Adaptive subgradient methods for online learning and stochastic optimization. In: COLT (2010)
3. Huang, P.S., He, X., Gao, J., Deng, L., Acero, A., Heck, L.P.: Learning deep structured semantic models for web search using clickthrough data. In: CIKM (2013)
4. Palangi, H., et al.: Semantic modelling with long-short-term memory for information retrieval. arXiv preprint arXiv:1412.6629 (2014)
5. Pang, L., Lan, Y., Guo, J., Xu, J., Wan, S., Cheng, X.: Text matching as image recognition. In: AAAI (2016)
6. Pennington, J., Socher, R., Manning, C.D.: Glove: global vectors for word representation. In: EMNLP (2014)
7. Shen, Y., He, X., Gao, J., Deng, L., Mesnil, G.: Learning semantic representations using convolutional neural networks for web search. In: WWW (2014)
8. Silver, D., et al.: Mastering the game of go with deep neural networks and tree search. Nature **529**(7587), 484–489 (2016)
9. Silver, D., et al.: Mastering the game of go without human knowledge. Nature **550**(7676), 354–359 (2017)
10. Sutton, R.S., Barto, A.G., Klopf, H.: Reinforcement Learning: An Introduction, 2nd edn (2016, in progress)
11. Wan, S., Lan, Y., Xu, J., Guo, J., Pang, L., Cheng, X.: Match-SRNN: modeling the recursive matching structure with spatial RNN. In: IJCAI (2016)

Classification

Music Mood Classification Based on Lifelog

Haoyue Tong, Min Zhang$^{(\boxtimes)}$, Pouneh Soleimaninejadian, Qianfan Zhang,
Kailu Wu, Yiqun Liu, and Shaoping Ma

Department of Computer Science and Technology,
Beijing National Research Center for Information Science and Technologys,
Tsinghua University, Beijing 100084, China
tong-14@mails.tsinghua.edu.cn, z-m@tsinghua.edu.cn

Abstract. Music mood analysis is crucial for music applications that involve search and recommendation. At present, the classification of music mood mainly depends on manual annotation or music information from audio and lyrics. However, manual annotation requires a large number of users, and the process of acquiring and processing audio or lyric information is complicated. Therefore, we need a new and simple method to analyze music mood. As music mood itself is specific psychological feelings caused by various music elements acting together on the user's hearing, in this paper we try to use information related to users, like their physiology and activities information. The development of wearable devices provides us with the opportunity to record user's lifelog. The experiment results suggest that the classification method based on user information can effectively identify music mood. By integrating with the classification method based on music information, the recognition effect can be further improved.

Keywords: Music mood classification · Lifelog · Ensemble learning

1 Introduction

Mood contained in music is an important part of a song. In the real world, mood categorization of music is crucial for music websites and applications. As users have the demand to search for music based on its mood, carriers need to know music mood so that they can build music indexes, in addition, carriers can also recommend music according to this. At present, mood classification methods on major music websites and applications are mainly based on tags given by users. Generally speaking, on the basis of human cognitive resonance, it is reasonable to use users tags as mood category label of music.

In order to analyze music mood, many classification models have been proposed. Considering that music mood is usually expressed by elements such as

This work is supported by Natural Science Foundation of China (Grant No. 61672311, 61532011).

timbre, tone, rhythm, melody and lyrics, these methods are mainly based on audio information extracted from music [1] or information about lyrics [2]. While music mood is the experience and feeling which conveys to the user, actually there is few research on music mood classification through the reaction of users.

With regard to the record of user conditions, lifelog has attracted more and more attention. Researchers have done a lot of research using lifelog. In terms of physical health, lifelog can be used to monitor and analyze users exercise and sleep status. For mental health, lifelog can be used to analyze users emotion and personality characteristics [3]. However, the value of lifelog has not yet been fully exploited.

We believe that lifelog could help mood analysis of music. Intuitively, some aspects of lifelog information, such as activity information, can be directly related to music mood. For example, users tend to choose rhythmically strong music when they are exercising, and soothe quiet music before falling asleep. Furthermore, as human body is a sensor system, users always have different physiological reactions when listening to music with different emotions.

In this paper, we propose to combine lifelog with music mood. Users own information and environment information are used to perform mood classification of the music the user listens to. First, we study the various contents of lifelog and extract biometrics, user's activities, user related features, and context information as the features of our experiments. Then we construct a model based on these features. Finally, we carry out experiments and results show that lifelog is helpful to music mood detection.

Our main contribution is to use lifelog for music mood detection and get better performance when using features of both lifelog and music. This is a brand-new point of view in music research, succinctly proposed by us in NTCIR-13 Lifelog-2 Task [3], that has unearthed the connection between music and human beings. The lifelog data set we collected has already been made public.

2   Related Work

2.1   Research Based on Lifelog

In previous work, many psychology researchers have explored the relationship between user's physiological characteristics and emotions, including blood pressure [4], heart rate [5], and galvanic skin response (GSR) [6]. In addition, the user's activity information such as the number of steps is also taken as one of the important features [7]. In these experiments, subjects usually need to wear specialized equipment to collect data. Limited by the conditions, most of experiments in this period only explored the impact of one kind of lifelog and the data volume was limited.

With the development of personal lifelog [14], more detailed user lifelog data is put into experiments [12]. The object of analysis, at the same time, is not confined to user emotions [8–11]. In addition to the user's emotions, Soleimanineja-dian et al. [3] analyzed user's personality characteristics, selecting gender, moody, optimistic, heart rate stability, room tidy and room decorative as features.

But so far, most of the researches using lifelog have analyzed characteristics of users, and few have attempted to analyze objects in other fields. Our research distinctively tries to combine user with music to analyze the emotion of music.

2.2 Methods of Music Classification

The analysis of music mood based on audio features is an important issue in the field of signal processing. Bai et al. [1] classified music mood classification as a problem in cognitive computing. They extracted more than 500 dimensions of features such as loudness from music and tried fuzzy KNN (FKNN), linear discriminant analysis (LDA) and some other classification algorithms.

There are also some studies that extract traits only for lyrics [15]. Corona et al. [2] extracted features from lyrics based on vector space models. Li et al. [16] used bag-of-character instead of bag-of-word to make full use of the information in Chinese lyrics and chose Deep Belief Network for training.

In addition, more researches have integrated multiple aspects of features. Xue et al. [17] proposed a multimodal approach to do music mood classification including audio and lyric information. Xiong et al. [18] introduced a generative multimodal method based on the correlation of audio and lyric information. Choi et al. [19] combined lyrics and user interpretations of lyrics to categorize music subjects.

It can be seen that most of the music analysis, whether for music mood or music subject, is to extract information from audio or lyrics. Although expressed through timbre, tone, rhythm, melody, lyrics and other elements, music mood is actually subjective feeling of the person who listens to it. Therefore, we creatively propose to analyze the mood of music from the conditions of the users.

3 Lifelog Based Music Mood Detection

Whether from a direct or indirect perspective, lifelog is helpful to the analysis of music mood. We treat it as a classification problem, and our research aims to classify music into specific mood categories. Note that in this article we do not emphasize the improvement and contribution to the model. As the first work to link music and lifelog information, in the current phase, our main purpose is to verify whether such an idea is feasible and whether the features we select are helpful to mood analysis of music. We believe if general models can work, the idea is feasible, and next step can be to improve specific model for better results.

3.1 Features

Through collection and screening of user life records, four types of features are extracted: users activities, biometrics, user related features, and context.

Users Activities. When he/she is listening to the music do correlated with the mood of the music. **The type and the intensity of the activity** that the user is performing often influence the user's choice of music. Besides, **the number of steps** the user takes while listening to music is also selected.

Biometrics. On the other hand, the mood of the music affects the users feelings. Intense music, for example, can make users more excited, heart rate faster and body temperature higher. Hence biometrics are other useful factors. The features selected include **calories, galvanic skin response (GSR), heart rate** and **skin temperature**.

User Related Features. As different users tend to have different habits and favor, we need to pay full attention to the differences among users. In addition, different users may not feel the same when they hear the same emotional song. Therefore, user-related features are classified into a separate category, including the **user's id** and the **user's gender**.

Context (Environment). In addition to factors directly related to the users, environmental factors may also have an impact on the user's choice of music. For example, there may be a pattern at what point in the day a user listens to certain kind of songs. At present, **timestamp** is the only feature in this category.

In order to make comparative study, we also obtain audio information and music metadata which are often used in traditional music classification experiments. This features have been proven to have a significant effect on the music mood classification.

We summarize lifelog and music features and their description in Table 1.

3.2 Model Based on 2D Thayer Mood Detection

An important model in mood analysis is Thayer's two-dimensional mood model [20], which has two axes of valence and arousal (energy). The valence dimension indicates a change of emotion from negative to positive, while the arousal dimension describes a change in energy from slight to strong.

When describing the emotions of a song, users usually analyzes from these two angles. However it is less likely to give specific values, but give some descriptive words. In fact we can map these words to the two-dimensional plane that matches the Thayer model. The music application Gracenote has done this work and they give the coordinates of the 25 kinds of music mood on the two-dimensional axis, as shown in Fig. 1.

As can be seen from the figure, high valence and high arousal in the upper right corner corresponds to the most positive and most energetic music mood Excited. Similarly, other emotions can also be found on the axes, depending on the valence and arousal.

3.3 Model Using Lifelog

Based on Thayers 2D mood model, the music mood detection problem can be taken as a classification problem. We propose two strategies to deal with this problem.

Table 1. Lifelog and music features set for music mood detection

Category	Feature	Description
Users activities	activity type	The type of the activity that the user is performing
	activity intensity	The intensity of activity that the user is performing
	steps	The number of steps the user takes
Biometrics	calories	The calories consumed by the user
	GSR	The response to changes or stimuli of the user's skin
	heart rate	The number of times the user's heart beats
	temperature	The body temperature of the user
User related features	user id	The number of the user
	user gender	The gender of the user
Context	timestamp	The time when the user data is recorded
Audio	valence	The more active the music, the higher the eigenvalue
	energy	The bigger the music, the higher the eigenvalue
	loudness	Mean loudness value of music
	tempo	The faster the rhythm, the higher the eigenvalue
	liveness	The more likely it is that the audiences voice is present in the music, the higher the eigenvalue
	danceability	The more suitable for dancing, the higher the eigenvalue
	key	The pitch of the music, 0 = C, 1 = C/D 2 = D...
Metadata	genre	The style of the music
	artist_era	The era of the artist
	artist_origin	The area of the artist
	artist_type	The membership of artists

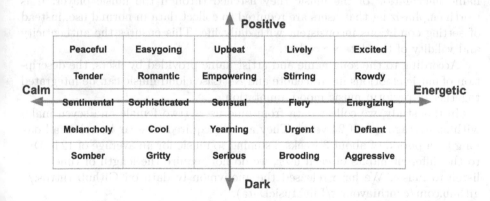

Fig. 1. Music mood labels on Thayers 2D mood model.

One-Step Detection. The first one is to directly classify music into 25 categories.

Two-Step Detection. The second is to build two classifiers for the two dimensions of valence and arousal respectively, and then combine these two results to obtain the final mood.

For the first strategy, because of the large number of classes, more data are needed to get good results. In the second approach, as the problem is divided in two stages, each classifier is relatively simple. But the error will be accumulated. Comparative experiment results are shown in Sect. 5.2. Our experimental design has no special requirements for classifiers, in other words, classifier is not the focus in our study. More complex models will be studied further in the future.

4   Lifelog Data Collection

As mentioned in Sect. 3.1, lifelog to be collected contains four pieces of information. Users are asked to provide basic personal information and wear smart bracelets to record biometrics, activity intensity and context information. We use the Gadgetbridge, an open source application (https://github.com/Freeyourgadget/Gadgetbridge), to synchronize data in the bracelet. This application currently only supports Android and does not support other systems. With Gadgetbridge, the bracelet can be set up for whole-day heart rate measurement and the detection frequency can be set to once a minute.

In addition, users are asked to record the time of the start and end of listening to music and the ongoing activities while listening to music. In order to obtain information about the music, users are asked to collect the song name and artist name information of the music they listened through the music player. It is worth emphasizing that users are required to collect data in normal use, instead of setting conditions inconsistent with daily life. This ensures the authenticity and validity of the data we collect.

According to the song name and artist name provided by users, the description of music emotions in Gracenote and NetEase cloud music can be integrated together to perform music mood annotation.

In this study, we collect data from nine users (two female and seven male) with an average age of 24 years. They are all required to wear bracelets all day long for a period of about 3 weeks, listening to music for an average of 12 h. Due to the different habits of each user, we do not require the length of time users listen to music. We have released the anonymously data on Github (https://github.com/cynthiayoung/LifeMusicData).

5   Experiment

5.1   Settings

Our data comes from two sources. The first part is collected using the method mentioned in Sect. 4 (called LifeMusic) and the second part is obtained through the data released by the NTCIR evaluation task (called NTCIR). The NTCIR evaluation task released data from two users (including user1's 59-day lifelog and user2's 31-day lifelog). This data set basically provides the data that need to be collected in Sect. 4. At the same time, music information are also recorded. Unfortunately, because user2 does not have the habit of listening to music, only user1's data is available to us. Therefore, the data used in this study involve a total of 10 users, of which the LifeMusic data set contains information for 9 users, and the NTCIR data set contains information for 1 user.

As mentioned above in Sect. 3.2, 25 kinds of music mood are used as labels. The following chart shows the distribution of the labels (Fig. 2).

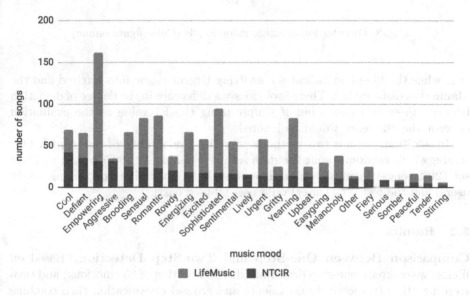

Fig. 2. Distribution of music mood labels (Color figure online)

The red part in the figure above shows the distribution of music mood tags in the LifeMusic dataset which is relatively uneven, and the blue part shows the distribution in the NTCIR dataset. Specifically, in the LifeMusic dataset there are some users who have obvious preferences when listening to music.

In this study, we use the the Manhattan distance between predicted result and actual mood on Thayer 2-D emotion coordinate plane instead of F1 score to measure the performance of our proposed model. When the coordinate of the prediction result is (x1, y1) and the actual music mood coordinate is (x2, y2), the Manhattan distance between them is:

$$distance = |x_1 - x_2| + |y_1 - y_2| \tag{1}$$

The following example gives a specific description of our evaluation function.

Figure 3 shows two cases of prediction and actual classification: the red line indicates classifying Urgent music into Yearning and the Manhattan distance

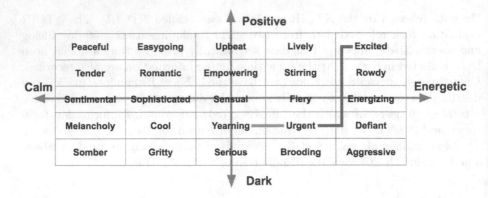

Fig. 3. Distribution of music mood labels (Color figure online)

is 1, while the blue line indicates classifying Urgent music into Excited and the Manhattan distance is 4. There is obviously a difference in the degree of deviation between these two cases, but if simply using the F1 value as the evaluation criteria, the difference would be ignored.

In addition, we use two methods to verify the validity of our model. One strategy is to randomly split the data set into training set (80% songs) and test set (20% songs) in music units. The other is to divide it into training set (9 users) and the test set (1 user) on a user-by-user basis.

5.2   Results

Comparison Between One-Step and Two-Step Detection. Based on lifelog, we compare one-step detection (direct detection of 25 emotions) and two-step detection (respectively do valence and arousal classification then combine these two results to obtain 25 emotions). Results are shown in Table 2.

Table 2. The comparison results among various strategies and classification algorithms

Strategy	Algorithm	Distance	Strategy	Algorithm	Distance
One-step	GBDT	2.21	Two-step	GBDT	1.82
One-step	Random Forest	2.24	Two-step	Random Forest	2.16
Two-step	XGBoost	1.90	Two-step	SVM	1.92
Two-step	AdaBoost	2.10	Two-step	KNN	2.34

As can be seen from the table, the two-step prediction results are better than those of the one-step for both GBDT and Random Forest algorithm. Although the method of two-step prediction will result in the accumulation of errors, but in fact, only one dimension with five species means fewer cases to consider.

Comparison Between Various Classification Algorithms. The algorithms we use for comparison include GBDT, XGBoost, AdaBoost, KNN, Random Forest, Decision Tree, Support Vector, etc. Some experimental results are also shown in Table 2. From the table, it can be seen that GBDT performs best. Furthermore, Boosting is very effective in this experiment as GBDT, XGBoost and AdaBoost are all evolved from Boosting.

Lifelog Features Analyses. Consistent with the categories presented in Sect. 3.1, we divide the lifelog information into four categories: users activities, biometrics, user related features and Context. We try to remove one of the features at a time in the experiment. After being removed, the greater the deviation of the results, the more important the feature is. Results are shown in Fig. 4.

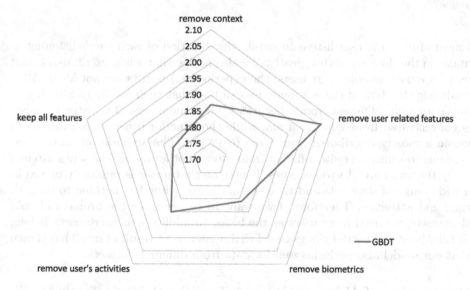

Fig. 4. The comparison results among using different lifelog features

It can be seen that the closer the distance from the center point is, the better the experimental result is, and the weaker the role of the non-used lifelog feature type corresponding to this dimension is. As a result, user related features are the most critical feature types in our experiment.

Cold Start Users Results. We try to set different user as cold start user. The so-called cold start user refers to the user's data as test set and the other user's data as training set. The results are shown Fig. 5.

The blue bar in the figure shows the length of time the user listens to music (the unit is minutes). The orange line indicates the result of the experiment when the user is a cold start user. Since there is no requirement and restriction on the

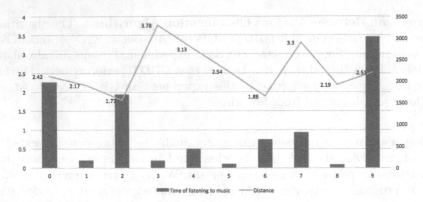

Fig. 5. The comparison results among using set different cold start user (Color figure online)

length of time the user listens to music, the duration of each user's listening to music in the data set varies greatly. We also notice that when user3, user4, and user7 are used as cold start users, the experimental results are not ideal. After analyzing the data of these 3 users, we can find that their habits of listening to music are quite different from other users. The time for user3 to listen to music is concentrated from 2 am to 4 am, while the time for other users to listen to music is mostly distributed from 8 am to 12 pm. The activities of user7 when listening to songs are quite different from those of other users. The songs listened to by the user4 are of various types. It may be that music is randomly played in a wide range of song lists and then the music recording has nothing to do with time and activities. Therefore, if we want to better solve the problem of cold start users, we need more users as the basis. In addition, as user0-user8 belong to LifeMusic and user9 belongs to NTCIR, experiment result of user9 has shown that our model also performs well for data from different datasets.

Combination of Using Lifelog and Traditional Music Features. We conduct comparative experiments with lifelog information and music information as features, and then integrate them together into a model. Results are as follows. From the table, it can be seen that although the experimental results of lifelog information are not as good as music information, the combination of them can further enhance the experimental results, indicating that the lifelog information has certain influence on the classification of music mood (Table 3).

Table 3. The comparison results among using lifelog, music and both (GBDT)

Feature	Lifelog	Music	Lifelog & music
Distance	1.82	1.14	1.12

5.3   Discussion

Music mood analysis using lifelog information is a brand-new attempt for mood analysis of music. In the classification of music mood, instead of relying entirely on music information, we can make more use of life information. Based on this, we can also make light-weight music recommendations to users afterwards.

From the above results, it can be seen that the features we propose are very helpful for the mood classification of music. After fusion with music information, we can get even better results.

At the same time, lifelog data collection is very convenient in the experiments we design. The smart sports bracelet we select can record user's heart rate, activity intensity, and number of steps per minute. Moreover, the sports bracelet is easy to carry, waterproof, and the battery lasts for about one week. As a result users will not be troubled when collecting data.

However, our research also has some limitations. In current experiment, very little information in lifelog has been used currently. Another important issue is that music without lifelog information cannot be analyzed. Therefore, classification based on lifelog could just be used as a supplement to the original music mood classification before large-scale application. What's more, the improvement of using lifelog features to enhance music information is not very marked and requires further study.

6   Conclusion and Future Work

In this study, we creatively propose a method for classifying musical mood from user's lifelog information and extract four types of lifelog features for experimentation. The results show that the user's lifelog features do contribute to the mood classification of music. Furthermore, we combine lifelog information with audio and music metadata to create a multi-modal model and achieve better results than using lifelog or audio and metadata alone.

Besides, we have released on Github the dataset we collect that contains lifelog and music information.

In the future we will continue to collect user data and conduct experiments on larger scale data. On this basis, more complex models and more features will be tried. In addition, music recommendation is also a focus of follow-up research. We can study the music mood that the user wants to listen to under specific activities and physiological conditions, and then automatically perform personalized music recommendations for the user.

References

1. Bai, J., et al.: Music emotions recognition by cognitive classification methodologies. In: 16th ICCI*CC (2017)
2. Corona, H., O'Mahony, M.P.: An exploration of mood classification in the million songs dataset. In: 12th SMC Conference. Maynooth University, Ireland (2015)

3. Soleimaninejadian, P., et al.: THIR2 at the NTCIR-13 lifelog-2 task: bridging technology and psychology through the lifelog personality, mood and sleep quality. In: NTCIR-13 (2017)
4. Davydov, D.M., Stewart, R., Ritchie, K., Chaudieu, I.: Depressed mood and blood pressure: the moderating effect of situation-specific arousal levels. Int. J. Psychophysiol. **85**(2), 212–223 (2012)
5. Johnston, D.W., Anastasiades, P.: The relationship between heart rate and mood in real life. J. Psychosom. Res. **34**(1), 21–27 (1990)
6. Vahey, R., Becerra, R.: Galvanic skin response in mood disorders: a critical review. IJP&PT **15**(2), 275–304 (2015)
7. Thayer, R.E., et al.: Amount of daily walking predicts energy, mood, personality, and health. In: Poster Presented at the APA, Washington DC (2005)
8. Kato, M.P., Liu, Y.: Overview of NTCIR-13. In: NTCIR-13 (2017)
9. Nam, Y., Shin, D., Shin, D.: Personal search system based on android using lifelog and machine learning. Pers. Ubiquit. Comput. **22**(1), 201–218 (2018)
10. Mafrur, R., Nugraha, I.G.D., Choi, D.: Modeling and discovering human behavior from smartphone sensing life-log data for identification purpose. HCIS **5**, 31 (2015)
11. Chung, C., Cook, J., Bales, E., Zia, J., Munson, S.A.: More than telemonitoring: health provider use and nonuse of life-log data in irritable bowel syndrome and weight management. J. Med. Internet Res. **17**(8), e203 (2015)
12. Maeda, M., Nomiya, H., Sakaue, S., Hochin, T., Nishizaki, Y.: Emotional video scene retrieval system for lifelog videos based on facial expression intensity. In: 18th SNPD, Kanazawa, pp. 551–556 (2017)
13. Byrne, D., Kelly, L., Jones, G.J.F.: Multiple multimodal mobile devices: lessons learned from engineering lifelog solutions. In: Software Design and Development: Concepts, Methodologies, Tools, and Applications (2014)
14. Jacquemard, T., Novitzky, P., OBrolchin, F., Smeaton, A.F., Gordijn, B.: Challenges and opportunities of lifelog technologies: a literature review and critical analysis. Sci. Eng. Ethics **20**(2), 379–409 (2014)
15. Kashyap, N., Choudhury, T., Chaudhary, D.K., Lal, R.: Mood based classification of music by analyzing lyrical data using text mining. In: ICMETE, Ghaziabad, pp. 287–292 (2016)
16. Li, J., Gao, S., Han, N., Fang Z., Liao, J.: Music mood classification via deep belief network. In: ICDMW, Atlantic City, New Jersey, pp. 1241–1245 (2015)
17. Xue, H., Xue, L., Su, F.: Multimodal music mood classification by fusion of audio and lyrics. In: He, X., Luo, S., Tao, D., Xu, C., Yang, J., Hasan, M.A. (eds.) MMM 2015. LNCS, vol. 8936, pp. 26–37. Springer, Cham (2015). https://doi.org/10.1007/978-3-319-14442-9_3
18. Xiong, Y., Su, F., Wang, Q.: Automatic music mood classification by learning cross-media relevance between audio and lyrics. In: ICME, pp. 961–966 (2017)
19. Choi, K., Lee, J.H., Hu, X., Downie, J.S.: Music subject classification based on lyrics and user interpretations. In: 79th ASIS&T (2016)
20. Russell, J.A.: A circumplex model of affect. J. Pers. Soc. Psychol. **39**(6), 1161–1178 (1980)
21. Wu, Y., Chang, E.Y., Chang, K.C., Smith, J.R.: Optimal multimodal fusion for multimedia data analysis. In: 12th ACM-MM, New York, pp. 572–579 (2004)

Capsule-Based Bidirectional Gated Recurrent Unit Networks for Question Target Classification

Shi Chen[1], Bing Zheng[2], and Tianyong Hao[3(✉)]

[1] School of Information Science and Technology,
Guangdong University of Foreign Studies, Guangzhou, China
AsherShiChen@outlook.com
[2] School of Informatics, University of Edinburgh, Edinburgh, UK
B.Zheng-1@sms.ed.ac.uk
[3] School of Computer Science, South China Normal University,
Guangzhou, China
haoty@126.com

Abstract. Question target classification, also known as answer type classification, turns out to be a key step for a high-performance QA system. The task has drawn more and more attention from academia and industry. This work proposes an artificial neural network architecture to classify questions according to their answer types. This architecture integrates several models. First, an embedding layer is trained using word2vec, an unsupervised language model, with a Chinese corpus from Wikipedia. Then, a bidirectional gated recurrent unit was concatenated on top of the embedding layer. A capsule layer and a softmax classifier is stacked on the architecture subsequently. Based on a SMP2017 dataset which contains over 3000 labeled Chinese questions, our method achieves an accuracy of 93.85% and F1-score of 93.68%, outperforming 6 baseline methods and demonstrating its effectiveness on question target classification.

Keywords: Question target classification · Capsule · Gated recurrent unit

1 Introduction

In open-domain Question Answering (QA) systems, the classification of users' natural language questions with respect to predefined answer types is an essential step in order to narrow down the scope of answer retrieval. This classification process, known as Question Target Classification (QTC) or answer type classification [1], can significantly promote the performance of QA systems by filtering out irrelevant answer candidates [2–4]. As addressed by Moldovan et al. [5], up to 36.4% of errors of a QA system can be avoided by performing accurate question target classification.

Despite the great success of statistical models on English QTC, researches focusing on Chinese datasets are relatively fewer. One possible reason is the lack of large scale publicly available annotation dataset. The National Conference of Social Media Processing (SMP) 2017 released a Chinese question classification dataset for a shared task

© Springer Nature Switzerland AG 2018
S. Zhang et al. (Eds.): CCIR 2018, LNCS 11168, pp. 67–77, 2018.
https://doi.org/10.1007/978-3-030-01012-6_6

-The Evaluation of Chinese Human-Computer Dialogue Technology (ECDT), which contains over 3000 manually labeled questions, giving researchers the opportunity to conduct practical experiments on the Chinese QTC research task.

To that end, this work proposes an artificial neural network architecture to classify question targets. First, we pre-train vectors using word2vec, an unsupervised language model, with a Chinese corpus from Wikipedia. The word vectors converted from questions are used in an embedding layer. Then, a kind of RNN, bidirectional Gated Recurrent Unit networks [6] was concatenated on top of the embedding layer. After that, a layer of capsules is applied. Capsules are analogous to perceptrons and their input is a set of vectors with a weight matrix, while output is a vector instead of scalar. Capsules is able to catch richer features from its input and produce more informative output. Finally, a softmax classifier is adopted in the output layer to determine the type of the input questions. Using the combination of these models, we have achieved an accuracy of 93.85% and F1-score of 93.68% on the SMP 2017 ECDT dataset [7].

The main contribution of this work lies on: (1) building a new neural network-based architecture for Chinese question target classification task, (2) exploring the effectiveness of capsule networks in model combination and proposing a new dynamic routing algorithm, (3) evaluating our model by comparing with baseline methods.

The rest of the paper is organized as follows: Sect. 2 introduces related work on question target classification. Section 3 describes the details of our method. Section 4 demonstrates dataset, evaluation metrics, and evaluation results, while Sect. 5 concludes the paper.

2   Related Work

The existing research work on question target classification can be divided into three strategies: rule-based, learning-based, and the combination of multiple strategies.

The rule-based strategy begins with detecting interrogative words, such as 'When', 'Where', 'Who', 'How' etc., which imply question type to some extent. For instance, a question starts with 'When' can be categorized in a type regarding time. However, questions that begins with interrogative word like 'How' and 'What' needs more complex information to confirm their answer type. Moreover, a question in QA systems unnecessarily contain an interrogative, e.g., 'Find the birthplace of Michael Jackson' [8, 9]. Therefore, rules are built to extract extra information except interrogative words, but the information is hard to find because of the complexity of natural language. Although rule-based strategy may achieve high performance, a great amount manual work is required to derive rules. Moreover, the rules are data-specific and are barely scalable, so they may not be easily applied to different types of question data [10].

There are some researches using the learning-based strategy. Li and Roth [2] applied a Sparse Network of Winnow (SNoW) to the question target classification problem. According to the definition of 6 coarse question classes and 50 fine classes, they trained hierarchical classifiers which mapped questions into coarse classes and then into fine classes. The method used a handful of features including POS tags and word conjunctions. Zhang and Lee [11] used support vector machine (SVM) with a kernel function based on syntactic trees. Henceforth, some works on question classification were tackling the self-defined advanced kernel of SVM [12–14]. Maximum Entropy models (ME), also known as Log Linear models, aroused interest of researchers such as Berger [15], as ME models calculated question's probability of falling into each class. Later promotions based on ME could mainly be attributed to better feature extraction. For example, Le Nguyen [16] parsed questions into trees and treated the problem as a tree classification problem, and subtrees were treated as features of ME model. Artificial-neural-network-based learning methods have become popular after they demonstrated their power in a wide range of problems. Kim [17] explored 4 variations of convolutional neural network (CNN) on the problem of question target classification on 7 datasets, and found CNN performed best on 4 of them compared with 14 reports shortly before 2014. Li et al. [18] proposed a context-aware model which combined a neural-network-based word embedding model word2vec and a classifier Probabilistic Graphical Models (PGM). Recent works [19, 20] showed capsule network's effectiveness on text classification tasks such as sentiment analysis, though capsule network by Sabour et al. [21] were originally designed for computer vision tasks.

Some hybrid strategies are proved to be successful on the question target classification problem. Huang et al. [22] used head word (a single word denoting a question's object) as a feature for SVM and ME classifiers and achieved an accuracy of 93.6% on coarse-grained classes and 89.0% on fine-grained classes on TREC dataset. In case a question started with interrogative word like 'when', 'where' or 'why', it was directly classified in terms of rules. Otherwise, 5 features including head word were extracted for SVM or ME classifier. Silva et al. [23] constructed question patterns to match an input question and thus determine its type. If no matched, the question was further parsed to extract its head word based on a refined rule-based parse tree method. Then, utilizing the information of head word with the help of WordNet (a dataset that can find hypernym of words) and disambiguation strategies, the question was mapped into its category. Furthermore, if the head word could not be extracted in the previous step, an SVM classifier was finally applied to determine the question type. Xie et al. [24] used word2vec model to automatically cluster semantically similar words to tackle the low coverage problem of WordNet. In this work, one of four question target word (QTW) extracting strategies was chosen concerning the dependency structure of input question. Then, the extracted QTW was expanded into question target feature using pretrained word2vec model and somehow used to classify the question. Consequently, Silva and Xie respectively achieved state-of-art accuracy on UIUC 5500 dataset using a combination of rule and learning-based strategies.

Although previous works have shown that methods performed decently on English question target classification task, most of them have not focused on Chinese questions. In addition, the popular artificial neural networks are also seldom being used for the task, partially due to the characteristics that questions are usually short and lack of context. This encouraged us to explore whether we can achieve higher performance with more targeted architecture of artificial neural networks on Chinese question target classification task.

3   The Capsule-Based Bi-GRU Architecture

We propose an architecture to deal with Chinese question target classification by combining Bidirectional-GRU(Bi-GRU), one of bidirectional RNN, with a capsule structure. As show in Fig. 1, the architecture contains six layers: (1) Input layer: input the question sentence and split it into characters; (2) Embedding layer: map each character to a distributed vector; (3) Bi-GRU layer: apply bidirectional GRU networks, to obtain features and information from the embedding layer; (4) Capsule layer: using the collection of vectors to replace the input and output of traditional multi-layer perceptron. Each vector represents a feature to express more abundant information. The layer updates its parameters with a dynamic routing algorithm. (5) Output layer: flatten the output of capsule layer and deliver it to the dense layer, where a softmax function is used to predict the classification.

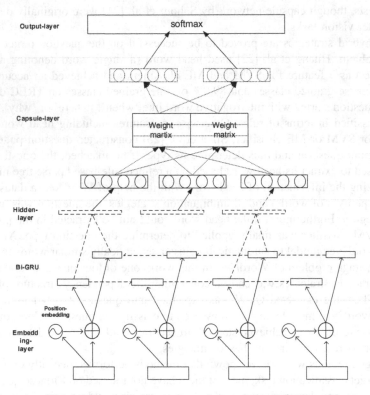

Fig. 1. The architecture of our proposed Cap-BGRU model

The details of the Bi-GRU layer and the Capsule layers, as the core of the architecture, are presented in Sects. 3.1 and 3.2, respectively. The embedding layers and the output layers are presented as follows.

In the embedding layer, after received a tokenized sentence from the input layer, we use a matrix to represent the sentence, the matrix rows are character vectors that represent each character in the sentence. The character vector comes from a corresponding pre-trained character vectors containing character relations. In this paper, we use CBOW [25] based on negative sampling to train the character vectors. The dimensionality of each character is fixed and predefined. By denoting the dimensionality and sentence numbers as d and s, the dimensionality of the sentence matrix thus is $s \times d$. Besides, because of the importance of sequence information for NLP tasks, we introduce a position embedding in the inspiration of Vaswani et al. [26] and represent the relative and absolute position of the character vector in sequential way.

Since the dimension of the position embedding is the same as the character vectors, we combine these two vectors as the output of embedding layer.

In the output layer, we use a softmax classifier to predict label $\hat{y}$ from a discrete set of classes Y for a sentence s. The classifier takes the flatten sequence f^* form capsule layer as input:

$$\hat{p}(y|S) = softmax\left(W^{(s)}f^* + b^{(s)}\right) \tag{1}$$

$$\hat{y} = \arg \max_{y} \hat{p}(y|s) \tag{2}$$

The loss function is the categorical cross entropy:

$$J(\theta) = -\frac{1}{N}\sum_{n=1}^{N} [y_n \log \hat{y}_n + (1 - y_n) \log(1 - \hat{y}_n)] \tag{3}$$

where y_i is truth label, $\hat{y}_i$ is the estimated probability for each class by the softmax function, and N is the number of target classes.

Different from the capsule model proposed by Zhao et al. [19], our model accesses the capsule layer after the RNN layer while their model is the combination of convolutional neural networks and different kinds of capsule networks. In addition, our model is different from another RNN-Capsule model proposed by Wang et al. [20], in which each emotion category is modeled as a capsule and an attention mechanism is built for capsule representation.

3.1 Bidirectional GRU Network

Proposed by Cho et al. [6], Gated Recurrent Unit (GRU) is a kind of Recurrent Neural Network. Like other RNNs, GRU uses its internal state (memory) to process sequences of inputs. As a variation of LSTM [27], GRU achieves a comparative performance but is simpler and faster in many tasks. It contains two gates: reset and update gates. The reset gate determines how to combine new input with previous memory, while the update gate defines how much of the previous memory to keep around.

The architecture of GRU is composited from four components. One input vector x_t with corresponding weight parameter matrix and vector, one update gate vector z_t with corresponding weight parameter matrix and vector W_z, U_z, b_z, one reset gate vector with corresponding weight parameter matrix and vector W_r, U_r, b_r, one output vector h_t with corresponding weight parameter matrix and vector W_h, U_h, b_h. The h_{t-1} is generated from the previous hidden state with the same structure to current state. The σ_r represents that rectified linear unit function is used as activation function. Denoting operator $\circ$ as the Hadamard product, the equations are demonstrated as follows:

$$z_t = \sigma(W_z x_t + U_z h_{t-1} + b_z) \tag{4}$$

$$r_t = \sigma(W_r x_t + U_r h_{t-1} + b_r) \tag{5}$$

$$h_t = (1 - z_t) \circ h_{t-1} + z_t \circ \sigma_r(W_h x_t + U_h(r_t \circ h_{t-1}) + b_h) \tag{6}$$

Therefore, the hidden unit depends on previous hidden units, reset gate and update gate. Those units that learn to capture short-term dependencies tend to have reset gates that are frequently active, but those that capture longer-term dependencies update gates that are mostly active by Cho et al. [6]. From the sequence point of view, standard GRU architectures are usually generated each unit by the past context information. Thus we can make use of both past and future context information.

In this work, bidirectional Gated Recurrent Unit (Bi-GRU) is utilized. The Bi-GRU network contains two sub-networks, one for positive time direction (forward states), and another for negative time direction (backward states). As shown in Fig. 1, the output of the ith character is the concatenation of the two-direction network's i^{th} hidden state output.

3.2   Capsule Structure

Capsule networks, a kind of neuron network mechanism, perform complicated internal computations on their inputs and then encapsulate the results of these computations into a small vector of highly informative outputs. The capsules can be divided into two types, lower-level and higher-level capsules. Each capsule represents some kinds of features. The higher-level capsules are decided by lower level capsules and a corresponding weight matrix.

We use a dynamic routing algorithm to compute the weight matrix and capsules. In order to make the output vector of a capsule to represent the probability that represented the below capsules and weight a current capsule's importance, we use a new non-linear function "*e_squash*", which is different from squash function, to shrink the vectors to more appropriate length. The corresponding equation is as follows:

$$v_j = \frac{\|s_j\|^2}{e^{-2} + \|s_j\|^2} \frac{s_j}{\|s_j\|} \tag{7}$$

where vector s_j is the total input of capsule j and v_j is its output vector. v_j comes from the computation of the first layer capsules. It is a sum of the product of c_{ij} and $\hat{u}_{j|i}$. $\hat{u}_{j|i}$ is produced by a shared weight matrix W_j, which multiplying the output u_i of the lower level capsules.

$$s_j = \sum_i c_{ij}\hat{u}_{j|i}, \hat{u}_{j|i} = W_j u_i \tag{8}$$

The procedure of our new dynamic routing algorithm in this architecture is summarized as follows:

Algorithm 1. Dynamic Routing

---

procedure Routing $(\hat{u}_j, r, l)$

 for all capsule i in layer l and capsule j in layer $(l + 1)$:$\boldsymbol{b}_i \leftarrow 0$

 for r iterations **do**

  for all capsule i in layer l: $\boldsymbol{c}_i \leftarrow softmax(\mathbf{b}_i)$

  for all capsule j in layer $(l + 1)$: $\boldsymbol{v}_j \leftarrow e\_squash(\sum_i c_{ij}\hat{u}_{j|i})$

  for all capsule i in layer l and capsule j in layer $(l + 1)$: $b_{ij} \leftarrow \hat{u}_{j|i} \cdot \boldsymbol{v}_j$

 return $\boldsymbol{v}_j$

In the algorithm, we initialize the capsule j with average values of vector $\hat{u}_{j|i}$. This process is different from the commonly used algorithm by Sabour et al. [21]. In addition, we replace the default "*squash*" function with a "*e_squash*" function. Moreover, in the final step of iterations, we refresh the value of b_{ij} by multiplying the output vector $\hat{u}_{j|i}$ from capsule i and the output vector from capsule j.

4 Experiment and Results

4.1 Evaluation Setup

In order to evaluate the effectiveness of our proposed method, the SMP2017 ECDT task 1 data [7], which is a publicly available dataset provided by iFLYTEK Corporation, was used. The dataset contains 3736 manually annotated Chinese questions and each question is manually labeled and assigned with a question target category. There are two first-level categories: chit-chat and task-oriented dialogue. Meanwhile, the task-oriented dialogue category further contains 30 second-level sub-categories. In addition, the dataset contains three sub-datasets: train, develop and test data. In this experiment, we used the chit-chat category and all of the 30 sub-categories as question target categories. The statistical characteristics of the dataset are shown in Table 1.

We trained 200-dimensional character vectors with dumps from latest Wikimedia documents using continuous bag of word (CBOW) based on negative sampling [28]. The dimension of output vectors encoded by Bi-GRU was setting to 600. The dropout 0.2 and 0.1 was applied in Bi-GRU layer and capsule layer, respectively. We used the Adadelta optimization algorithm to train our architecture. To optimize loss, we used 6 iterations routing for all datasets. The architecture was trained with a batch size of 64 examples.

In this experiment, we used several state-of-the-art baseline methods including SVM, CNN, LSTM, GRU, Bi-LSTM, Bi-GRU. For SVM, we used a linear kernel and transformed input data into vectors with TF-IDF weights. In other neural network

Table 1. The statistical characteristics of the dataset SMP2017ECDT-DATA

Statistics	Count
# of train questions	3026
# of development questions	336
# of test questions	374
# of total questions	3736
# of total characters	31318
Maximum # of question characters	65
Minimum # of question characters	2
Standard deviation of question characters	8.38276

models, we used the same embedding vectors and all of them with 300 output units. Other parameters were set as commonly used default values.

4.2 The Results

To evaluate the stability and the applicability of our method, an experiment was conducted by randomly dividing the dataset into 6 sub sets with different sizes. These sets containing 1000, 1500, 2000, 2500, 3000 and 3500 questions, respectively. After that, each of the neural network models including Bi-GRU, GRU and LSTM as baseline methods as well as our Cap-BGRU model were ran on every sub datasets. The training data in each sub dataset took 90% proportion. To prevent overfit problem and acquire the best error estimation, we chose tenfold cross-validation to train these models. The experiment results using F1-sorce metric are shown in Table 2. From the result, all the models achieved relative increasing performance. Our Cap-BGRU obtained higher performance using lower number of training questions. With the increasing size of question data, our Cap-BGRU was more effective to achieve optimized performance considering training dataset was usually limited.

Table 2. The classification performance with different numbers of questions

# questions	Bi-GRU	GRU	LSTM	Cap-BGRU
1000	76.00%	76.00%	77.60%	85.65%
1500	85.99%	84.66%	86.62%	89.03%
2000	86.49%	85.99%	86.70%	91.14%
2500	86.79%	87.59%	89.65%	89.51%
3000	85.29%	88.00%	87.10%	90.01%
3500	90.26%	91.45%	88.37%	92.52%

In addition, we further compared the performance of our method to all the six baseline methods. The comparison was conducted on the whole SMP2017 ECDT dataset with tenfold cross validation as well. The total 3736 questions thus were randomly divided into ten even subsets. In each round, nine of the ten subsets were used for training and the remaining one was used for testing. The evaluation measures are widely-used Accuracy, Precision, Recall and F1-Measure in information retrieval area. The result is shown in Table 3.

Table 3. The performance comparison of our Cap-BGRU model with six baseline methods on the whole dataset

Method	Accuracy	Precision	Recall	F1-sorce
SVM	73.46%	85.83%	73.46%	75.18%
CNN	71.39%	72.46%	71.12%	71.77%
LSTM	89.0%	90.86%	88.77%	89.49%
Bi-LSTM	89.30%	91.10%	88.77%	89.89%
GRU	90.37%	92.10%	90.10%	91.08%
Bi-GRU	89.30%	91.04%	88.23%	89.60%
Cap-BGRU	93.85%	94.55%	93.58%	94.05%

From the result, the popular Bi-LSTM model obtained an accuracy of 89.3%, a precision of 91.1%, a recall of 88.77%, and a F1-score of 89.89%. The performance was higher than LSTM but lower than that of GRU model. Our model obtained an accuracy of 93.85%, a precision of 94.55%, a recall of 93.58%, and a F1-score 93.68%, achieving the best performance among all the evaluation metrics. We identified that the Bi-GRU performed slightly worse than the GRU model. The results presented that our capsule based RNN model with a modified dynamic routing algorithm outperformed all the baseline neural network models. The use of capsule not only captured richer features from former layers but also output vectors which was more informative to the subsequent layers. Moreover, RNN architecture and position embedding were used to construct the hidden layer of the architecture. This architecture did not involve NLP tools or other linguistics resources except the Chinese Wikipedia corpus. Our method did not used lexical resource or dependency parser to get high-level features but still achieved higher performances.

We also explored the effect of the sequence of input text to the model performance. with the reversed sequence of input text, both GRU model and Bi-GRU model ran on the testing data with the same parameter settings. The results was show in Table 4. The performance of the two models slightly declined with the reversed text sequence, from 91.08% to 89.70% and from 89.6% to 89.24%, respectively. Therefore, there was a correlation between the RNN model and text sequence.

Table 4. The performance of the GRU model and Bi-GRU model using reversed text sequence

Method	Accuracy	Precision	Recall	F1-sorce
GRU	89.30%	90.16%	89.03%	89.70%
Bi-GRU	89.30%	90.59%	87.96%	89.24%

5   Conclusions

This paper proposed a neural-network-based architecture for Chinese question target classification. The architecture featured with a capsules-based bidirectional GRU model, which contains six layers. Using the publicly available SMP 2017 ECDT data as experiment dataset, we compared the architecture with six state-of-the-art baseline methods. The experiment results presented that the proposed model achieved the best performance among all the baselines. The result demonstrated the effectiveness of our proposed Cap-BGRU architecture on Chinese question target classification.

Acknowledgements. This work is supported by grants from National Natural Science Foundation of China (No. 61772146) and Guangzhou Science Technology and Innovation Commission (No. 201803010063).

References

1. Hao, T., Xie, W., Chen, C., Shen, Y.: Systematic comparison of question target classification taxonomies towards question answering. In: Zhang, X., Sun, M., Wang, Z., Huang, X. (eds.) CNCSMP 2015. CCIS, vol. 568, pp. 131–143. Springer, Singapore (2015). https://doi.org/10.1007/978-981-10-0080-5_12
2. Li, X., Roth, D.: Learning question classifiers. In: Proceedings of the 19th International Conference on Computational Linguistics. Association for Computational Linguistics, vol. 1, pp. 1–7 (2002)
3. Ittycheriah, A., Franz, M., Roukos, S.: IBM's statistical question answering system-TREC-10. In: TREC (2001)
4. Hovy, E., Gerber, L., Hermjakob, U., Lin, C.Y., Ravichandran, D.: Toward semantics-based answer pinpointing. In: Proceedings of the First International Conference on Human Language Technology Research. Association for Computational Linguistics, pp. 1–7 (2001)
5. Moldovan, D., Paşca, M., Harabagiu, S., Surdeanu, M.: Performance issues and error analysis in an open-domain question answering system. ACM Trans. Inf. Syst. (TOIS) **21**(2), 133–154 (2003)
6. Cho, K., et al.: Learning phrase representations using RNN encoder-decoder for statistical machine translation. arXiv preprint arXiv:1406.1078 (2014)
7. Zhang, W.N., Chen, Z., Che, W., Hu, G., Liu, T.: The first evaluation of chinese human-computer dialogue technology. arXiv preprint arXiv:1709.10217 (2017)
8. Hull, D.A.: Xerox TREC-8 question answering track report. In: TREC (1999)
9. Prager, J., Radev, D., Brown, E., Coden, A.: The use of predictive annotation for question answering in TREC8. In: Eighth Text REtrieval Conference. NIST Special Publication 500-246 (1999)
10. Li, X., Roth, D.: Learning question classifiers: the role of semantic information. Nat. Lang. Eng. **12**(3), 229–249 (2006)
11. Zhang, D., Lee, W.S.: Question classification using support vector machines. In: Proceedings of International ACM SIGIR Conference on Research and Development in Information Retrieval. ACM, pp. 26–32 (2003)
12. Moschitti, A., Quarteroni, S., Basili, R., Manandhar, S.: Exploiting syntactic and shallow semantic kernels for question answer classification. In: Proceedings of the 45th Annual Meeting of the Association of Computational Linguistics, pp. 776–783 (2007)

13. Pan, Y., Tang, Y., Lin, L., Luo, Y.: Question classification with semantic tree kernel. In: Proceedings of the 31st Annual International ACM SIGIR Conference on Research and Development in Information Retrieval. ACM, pp. 837–838 (2008)

14. Tomas, D., Giuliano, C.: A semi-supervised approach to question classification. In: ESANN (2009)

15. Berger, A.L., Pietra, V.J.D., Pietra, S.A.D.: A maximum entropy approach to natural language processing. Comput. Linguist. **22**(1), 39–71 (1996)

16. Le Nguyen, M., Tri, N.T., Shimazu, A.: Subtree mining for question classification problem. In: Proceedings of IJCAI, pp. 1695–1700 (2007)

17. Kim, Y.: Convolutional neural networks for sentence classification. arXiv preprint arXiv: 1408.5882 (2014)

18. Li, H., Wang, N., Hu, G., Yang, W.: PGM-WV: a context-aware hybrid model for heuristic and semantic question classification in question-answering system. In: 2017 International Conference on Progress in Informatics and Computing (PIC). IEEE, pp. 240–244 (2017)

19. Zhao, W., Ye, J., Yang, M., Lei, Z., Zhang, S., Zhao, Z.: Investigating capsule networks with dynamic routing for text classification. arXiv preprint arXiv:1804.00538 (2018)

20. Wang, Y., Sun, A., Han, J., Liu, Y., Zhu, X.: Sentiment analysis by capsules. In: Proceedings of the 2018 World Wide Web Conference on World Wide Web, pp. 1165–1174 (2018)

21. Sabour, S., Frosst, N., Hinton, G.E.: Dynamic routing between capsules. In: Advances in Neural Information Processing Systems, pp. 3859–3869 (2017)

22. Huang, Z., Thint, M., Qin, Z.: Question classification using head words and their hypernyms. In: Proceedings of the Conference on Empirical Methods in Natural Language Processing. Association for Computational Linguistics, pp. 927–936 (2008)

23. Silva, J., Coheur, L., Mendes, A.C., Wichert, A.: From symbolic to sub-symbolic information in question classification. Artif. Intell. Rev. **35**(2), 137–154 (2011)

24. Xie, W., Gao, D., Hao, T.: A feature extraction and expansion-based approach for question target identification and classification. In: Wen, J., Nie, J., Ruan, T., Liu, Y., Qian, T. (eds.) CCIR 2017. LNCS, vol. 10390, pp. 249–260. Springer, Cham (2017). https://doi.org/10. 1007/978-3-319-68699-8_20

25. Mikolov, T., Sutskever, I., Chen, K., Corrado, G.S., Dean, J.: Distributed representations of words and phrases and their compositionality. In: Advances in Neural Information Processing Systems, pp. 3111–3119 (2013)

26. Vaswani, A., et al.: Attention is all you need. In: Advances in Neural Information Processing Systems, pp. 6000–6010 (2017)

27. Hochreiter, S., Schmidhuber, J.: Long short-term memory. Neural Comput. **9**(8), 1735–1780 (1997)

28. Goldberg, Y., Levy, O.: Word2vec explained: deriving Mikolov et al.'s negative-sampling word-embedding method. arXiv preprint arXiv:1402.3722 (2014)

Question-Answering Aspect Classification with Multi-attention Representation

Hanqian Wu[1,2(✉)], Mumu Liu[1,2], Jingjing Wang[3], Jue Xie[4], and Shoushan Li[3]

[1] School of Computer Science and Engineering, Southeast University, Nanjing, China
hanqian@seu.edu.cn, liudoublemu@163.com
[2] Key Laboratory of Computer Network and Information Integration
of Ministry of Education, Southeast University, Nanjing, China
[3] NLP Lab, School of Computer Science and Technology, Soochow University,
Suzhou, China
djingwang@gmail.com
[4] Southeast University-Monash University Joint Graduate School, Suzhou, China

Abstract. In e-commerce platforms, the question-answering style reviews are emerging, which usually contains much aspect-related information about products. In this paper, Question-answering (QA) aspect classification is a new task that aims to identify the aspect category of a given QA text pair. According to characteristics of QA-style reviews, we draw up annotation guidelines and build a high-consistency annotated corpus for QA aspect classification. Then, we propose a recurrent neural network based on multi-attention representation to tackle this new task. Specifically, we firstly segment the answer text into clauses, and then leverage the multi-attention representation layer to match the question text with clauses inside answer text and generate multiple attention representations of the question text, which extends feature information of the question text. The experimental results demonstrate that our method for QA aspect classification, which is based on multi-attention representation, can make the most of useful information in answer texts and perform better than some strong baselines in QA aspect classification.

Keywords: Question answering · Aspect classification
Attention mechanism

1 Introduction

Recently, there appears a large number of user-generated question-answering (QA) reviews in various e-commerce platforms, such as Amazon, Taobao and Kaola. An example of QA-style reviews is shown in Fig. 1. By this novel form of reviewing, the potential customers can ask questions about certain product and others who have purchased the same item kindly answer these questions. Thus, this QA-style reviews are more reliable and convincing than traditional reviews written by any users. However, very few research has been conducted on aspect

© Springer Nature Switzerland AG 2018
S. Zhang et al. (Eds.): CCIR 2018, LNCS 11168, pp. 78–89, 2018.
https://doi.org/10.1007/978-3-030-01012-6_7

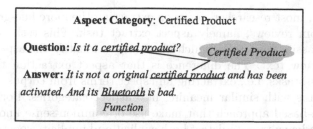

Aspect Category: Certified Product

Question: *Is it a <u>certified product</u>?*   *Certified Product*

Answer: *It is not a original <u>certified product</u> and has been activated. And its <u>Bluetooth</u> is bad.*
Function

Fig. 1. A translated example of QA-style reviews from an e-commerce platform.

classification of QA-style reviews which aims to identify the aspect category of a given QA text pair.

According to analysis of corpus, as shown in Fig. 1, we can find that **Question** only contains one aspect *certified product*, while there are two different aspects contained in different clauses in **Answer**, i.e., the clause *"It is not a original certified product and has been activated."* relating to the aspect *certified product* and the clause *"And its Bluetooth is bad"* relating to the aspect *function*. From the point view of consumers questioning, the expected answer should be only related to the aspect *certified product*. Thus, for this QA text pair, we should only consider how to identify the aspect *certified product* involved in **Answer**. Inspired by this, we can firstly segment *Answer* into clauses to make each clause contain only one aspect. Then we can leverage a attention-based layer to match **Question** with each clause in **Answer** and capture the most relevant information between **Question** and **Answer**.

In view of the above issue, we first draw up some specifically designed annotation guidelines and build a high-consistency corpus for our QA aspect classification task. On this basis, we propose a recurrent neural network based on multi-attention representation to solve this task. Specifically, we segment the answer text into different clauses, and then leverage the multi-attention representation layer to match the question text with each clause inside answer text and obtain multiple attention representations of the question text to extend the features of question text for classification. The empirical studies demonstrate that our proposed approach can make full use of the relevant information in the answer text to achieve better performance and outperform other baseline methods.

2   Related Work

Aspect classification, i.e., aspect category classification, can be treated with approaches applied to text classification, such as CNN [5], LSTM [11]. However, there is very few research with focus on aspect classification task. Toh et al. [12] leverage the sigmoidal feedforward network to train binary classifiers for aspect category classification. Xue et al. [15] perform joint learning with the two tasks aspect category classification and aspect term extraction based on neural networks.

In addition, most researchers focus on extracting the more fine-grained opinion targets from reviews, namely aspect extract task. This task is associated with aspect classification task, which also aims to identify the aspect category of a given review text. The difference is that aspect extraction task usually includes two subtasks, i.e., extracting all aspect terms from corpus and clustering aspect terms with similar meaning into aspect categories. Poria et al. [8] propose a rule-based approach that make use of common-sense knowledge and sentence dependency trees to detect both explicit and implicit aspects from opinionated texts. Rana et al. [9] mine sequential patterns from customer reviews and define rules on the basis of these patterns, and they then propose a two-fold rule-based method, in which the first fold extracts aspect associated with domain independent opinions and the second fold extracts aspects associated with domain dependent opinions. Conditional Random Fields (CRF) [6] method requires much manual annotation. Inspired by the approach [1], Shu et al. [10] propose a lifelong CRF model for aspect extraction which leverages the knowledge from many past domains to assist extraction for a new domain. Unsupervised methods are also applied to avoid reliance on annotated data. For instance, in recent years, Latent Dirichlet Allocation (LDA) and its variants have become the dominant unsupervised approach to aspect extraction [7]. However, LDA-based models need to estimate a distribution of topics for each document. To address the above challenge of the LDA-based methods, He et al. [4] propose a attention-based neural approach to emphasize aspect-related words to further improve the coherence of aspects.

Furthermore, as far as we know, our study takes the lead in the aspect classification task for QA text pairs, which distinctly differs from the above existing research.

3   Data Collection and Annotation

Our data are mainly form *"Asking All"* in Taobao[1], which is the most famous e-commerce platform in China. We extract 8,313 QA text pairs from the *electronic appliances* domain and manually annotate them. To ensure the high consistency of annotation, we draw up three aspect-related annotation guidelines (G1, G2 and G3) and assign two annotators to label each QA text pair in the form of a triple of *aspect term*, *aspect* and *polarity*. Note that all examples presented in this paper are translations of original Chinese texts.

G1: If we extract the aspect term from the question text, we will consider to annotate this QA text pair with a triple. According to the headword of an aspect term, the extraction fineness of aspect terms can be divided into the following situations.

(a) If the headword is a verb, the extraction fineness of the aspect term is the verb with a noun which follows closely.

---

[1] https://www.taobao.com/.

(b) If the headword is a noun, the extraction of the aspect term will conform to the principle of noun phrase maximization.

(c) If the headword is an adjective or adverb, the aspect term is the adjective or adverb.

G2: All aspects, i.e., aspect categories, are predefined based on the extraction of aspect terms in our corpus. For instance, aspect terms *"screen"* and *"signal"* extracted from question texts can be classified as the aspect category *"IO"*. Aspects are divided into two categories, one is domain-independent aspect, i.e., the aspect may appear in all domains, such as *weight, quality* and *appearance*. The other one is domain-dependent aspect, i.e., the aspect is specific to the particular domain, such as *performance, battery* and *IO* which only exist in *electronic appliances* domain.

G3: Once an aspect is mentioned in both the question and answer text, the next step is to decide the sentiment polarity of this aspect and then annotate it with a triple. Generally speaking, the sentiment polarity can be subdivided into *positive, negative* and *conflict* (a mix of both positive and negative) and *neutral* categories. We may come across the following cases during the processing of annotation,

(a) If the answer text expresses the objective evaluation about the aspect referred in the question text, the QA text pair is annotated as (*aspect term, aspect, neutral*). **E1** is an example of this category. In the question text, the aspect term is related to the aspect *appearance*, and the clause *"Its back is not very flat,"* objectively evaluates it, so we annotate this QA text pair with a triple of (*flat, appearance, neutral*).

E1: Q: Is the back of this phone flat?

A: Its back is not very flat, but it feels good. And its signal is too bad.

(b) If the answer text contains negative sentimental words related to the aspect in the question text, such as *"too bad"* and *"not good"*, the QA text pair is annotated as (*aspect term, aspect, negative*). **E2** is an example of this category. The aspect term *"choppy"* is related to the aspect *performance* involved in both the question and answer text, and the answer text expresses negative sentiment of it. Though the answer text also expresses the negative sentiment of camera pixels, the question text does not refer to pixels. Thus, this QA text pair is annotated as (*choppy, performance, negative*).

E2: Q: Is it choppy when you are playing games?

A: It fails to work well and is choppy. And the pixels of camera are low.

(c) If there exists sentimental expressions like *"great"* and *"good"* in the answer text and they are related to specific aspect in the question text, the QA text pair is annotated as (*aspect term, aspect, positive*). **E3** is an example of this category.

E3: Q: Does this phone peel off paint easily?

A: No, it does not and I have been using it for a long time. And its signal is good.

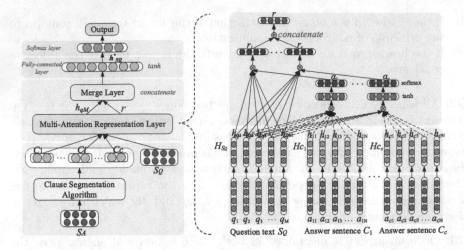

Fig. 2. The overall architecture of our approach.

(d) Given an aspect in the question text, if the answer text contains both positive and negative sentiment, the QA text pair is annotated as (*aspect term, aspect, conflict*). **E4** is an example of this category which expresses both positive and negative sentiment about the aspect term "*signal*" relating to the aspect *IO*.

E4: Q: How about its signal?
A: It has a good semaphore when using the mobile phone, but the connection of wireless network fails to work well. And it makes its memory card easily hot.

The Kappa consistency check value of the annotation is 0.81. To cope with the inconsistently annotated QA text pairs, an expert is assigned to proofread them. After annotation, we finally build a high-consistency corpus with annotated 2,566 QA text pairs which conform to the above annotation guidelines. In this paper, our goal is to identify the aspect category of a given QA text pair. And if this paper is accepted, we will release this high-consistency annotated corpus.

4   Our Approach

In this section, we firstly introduce the clause segmentation algorithm of the answer text, and then we propose a recurrent neural network based on multi-attention representation to identify the aspect category of a given QA text pair. Figure 2 shows the overall architecture of our proposed approach to the QA aspect classification task. We will describe our approach in detail in the following sections.

Algorithm 1: Clause Segmentation Algorithm

Input: Answer text $S_A = \{w_i \mid w_i \text{ is a word.}\}$;
V_p: Chinese punctuations set;
N_{min}: the minimum number of words in a clause;
N_{max}: the maximum number of clauses in the answer text
Output: All clauses (Stored in $C = \{ c_i \}$) mined from S_A that satisfy N_{min}
and N_{max}.

1 $C = \emptyset$;
2 $c_{temp} = null$; //the candidate clause
3 Segment S_A into n clauses $\{a_1, \cdots, a_n\}$ with V_p;
4 **for** $i = 1$; $i \leq |S_A| - 1$; $i{+}{=}1$ **do**
5   **if** $|C| \geq N_{max}$ **then**
6   break;
7   **end**
8   **if** $a_i.length > N_{min}$ **then**
9          $c_{temp} = a_i$;
10          $C = C \cup \{c_{temp}\}$;
11   **else**
12          $j = i + 1$;
13   **while** $j \leq |S_A| - 1$ **do**
14              $c_{temp} = a_i + a_j$;
15   **if** $c_{temp}.length \geq N_{min}$ **then**
16                  $C = C \cup \{c_{temp}\}$;
17   **else**
18                  $j{+}{=}1$;
19   **end**
20   **end**
21          $i = i + j$;
22   **end**
23 **end**

4.1 Clause Segmentation Algorithm

As described in Sect. 3, clauses inside answer text could contain different aspects in a QA text pair and only one clause is related to the annotated aspect. Thus, we segment the answer text into clauses to capture useful information contained in the answer text for classification.

The main idea of clause segmentation algorithm is to segment answer texts with Chinese punctuations as delimiters. For a clause in the answer text, we define the minimum number of words as N_{min}. And only when the length of one clause is larger than N_{min}, it is called a clause. Besides, we define the maximum number of clauses as N_{max} to determine the number of clauses required for the input of the neural network. Algorithm 1 describes the clause segmentation algorithm in detail.

4.2 Multi-attention Representation Layer

The core of our proposed approach is to capture the most relevant information between the question and answer text by leveraging the multi-attention representation layer and extend the feature representation of the question text to improve the performance of QA aspect classification. The right part in Fig. 2 depicts how to obtain the multiple attention representations of the question text in details.

For a given QA text pair, assume that the answer text S_A has been segmented into c clauses $\{c_1, \cdots, c_c\}$ and each clause contains N words. The vector representation $a_{ij} \in R^{d_w}$ denotes the j-th word of the i-th answer clause. The question text S_Q contains M words, and the vector representation $q_i \in R^{d_w}$ denotes the i-th word in the question text, where d_w represents the dimension of word embeddings in the question/answer text.

First, we encode the question text S_Q and the answer clause c_i with LSTM model [11,13], where $i \in [1,c]$, to obtain the hidden state matrix $H_{S_Q} = [h_{q1}, \cdots, h_{qM}]$ of S_Q and $H_{c_i} = [h_{i1}, \cdots, h_{iN}]$ of c_i by the following formulas,

$$H_{S_Q} = \mathrm{LSTM}(S_Q) \tag{1}$$

$$H_{c_i} = \mathrm{LSTM}(c_i) \tag{2}$$

where $H_{S_Q} \in R^{N_w \times d_h}$, $H_{c_i} \in R^{N_w \times d_h}$, N_w is the number of words in the question text or the answer clause, and d_h is the size of LSTM hidden layer.

Further, we compute the attention weight vector α_i between H_{S_Q} and H_{c_i} to capture the most relevant information relating to the annotated aspect between question sentence S_Q and answer clause c_i as follows,

$$M_i = \tanh(W_i \cdot (H_{S_Q}{}^T \cdot H_{c_i}) + b_i) \tag{3}$$

$$\alpha_i = \mathrm{softmax}(W_e{}^T \cdot M_i) \tag{4}$$

where $1 \leq i \leq C$, $M_i \in R^{N_w \times N_w}$, $\alpha_i \in R^{N_w}$, W_i and W_e are the weight matrices, b_i is the bias and $\cdot$ denotes the dot product between matrices.

Then, we obtain the attention representation $r_i \in R^{d_h}$ of the question text S_Q based on the weights, i.e.,

$$r_i = H_{S_Q} \cdot \alpha_i^T \tag{5}$$

The answer text is segmented into c clauses, so we can obtain the attention representation set $R = \{r_1, \cdots, r_i, \cdots, r_c\}$ of the question text S_Q where $|R|$ is c. And we concatenate these attention representations together into a new vector $r \in R^{d_h}$.

$$r = r_1 \oplus \cdots \oplus r_i \oplus \cdots \oplus r_c \tag{6}$$

Besides, according to the guidelines **G1** and **G2** in Sect. 3, QA aspect classification task mainly depends on the question text. Thus, the final feature representation $h^* \in R^{d_h}$ of S_Q is computed by concatenating r with the last hidden vector $h_{qM} \in R^{d_h}$ of the question text as follows,

$$h^* = \tanh(W_p r + W_x h_{qM}) \tag{7}$$

where W_p and W_x are the weight matrices.

Finally, a *softmax* layer is followed to obtain the conditional probability distribution:

$$y = \text{softmax}(W h^* + b) \tag{8}$$

where W and b are parameters for the *softmax* layer. On this basis, the label with the highest probability stands for the predicted aspect category of a QA text pair.

4.3 Model Training

Cross-entropy loss function is used to train our model end-to-end for classification. Given a set of training data S_{Q_t}, S_{A_t} and y_t, where S_{Q_t} is the t-th question text, S_{A_t} is the corresponding answer text, and y_t is the ground-truth aspect for a QA text pair (S_{Q_t}, S_{A_t}), if we represent this model as a black-box function $\phi(S_Q, S_A)$, whose output is a vector representing the probability of aspects, then the optimization goal of training is to minimize the loss function:

$$J(\theta) = -\sum_{t=1}^{N_s} \sum_{k=1}^{K} y_t^k \cdot \log \phi(S_{Q_t}, S_{A_t}) + \frac{l}{2}||\theta||_2^2 \tag{9}$$

where N_s is the number of training samples, K is the number of aspects for classification and l is a L_2 regularization to bias parameters.

In the equation above, we adopt *Adagrad* optimizer [2] to optimize parameters in our model, and initialize all the matrix and vector parameters with uniform distribution $[-\sqrt{6/(r+c')}, \sqrt{6/(r+c')}]$, where r and c' are rows and cols of the matrix respectively [3]. Besides, the dropout strategy is used in LSTM layer to avoid over-fitting.

5 Experimentation

5.1 Experimental Settings

- **Data Settings:** Due to the imbalance of distribution of data, aspect categories which contain less than 50 QA text pairs are omitted. Table 1 depicts the distribution of experimental data. Besides, we set aside 10% from the training data as the development data to tune learning algorithm parameters.
- **Word Representations:** Word embedding is used for feature representations of experimental data, which is pre-trained based on Skip-Gram model with Gensim [3] toolkit and 320 thousand QA text pairs collected from "*Asking All*" in Taobao.

Table 1. Data distribution in our experiment.

Aspect	Amount of QA text pairs
Performance	548
Battery	230
IO	908
Function	111
Quality	165
Certified product	370
Computation	95

- **Evaluation Metrics:** The evaluation metrics of performance are mainly *Accuracy* and *Macro-F1* (F) which is calculated by the formula $F = \frac{2PR}{P+R}$, where the overall precision P and recall R are the average of the precision/recall scores of all categories. Furthermore, we use t-test to assess the significance of the performance difference between two approaches [16].
- **Hyper-parameters:** In our experiment, the dimensions of word embeddings and LSTM hidden layers are set to be 100. The other hyper-parameters are tuned according to the development data. Specifically, the learning rate is 0.01 and the dropout rate is 0.4. And in our clause segmentation algorithm, the minimum number of words in a clause is 5 and the maximum number of clauses inside the answer text is 3. Besides, all out-of-vocabulary words are initialized by sampling from the uniform distribution $U(-0.01, 0.01)$.

5.2   Experimental Results

For a comprehensive analysis and comparison, we implement some baselines for QA aspect classification task to evaluate the performance of our proposed approach. And all approaches use the same word representations.

- **CNN(A):** This basic baseline approach proposed by Kim et al. [5] takes answer texts as input of CNN.
- **CNN(Q):** This basic baseline approach takes question texts as input of CNN.
- **CNN(Q+A):** This basic baseline approach takes the concatenation of question and answer texts as the input of CNN.
- **LSTM(A):** This is a baseline approach which puts answer texts into the input layer of LSTM proposed by Tang et al. [11].
- **LSTM(Q):** This is a baseline approach which puts question texts into the input layer of LSTM.
- **LSTM(Q+A):** This is a baseline approach which puts the concatenation of question and answer texts into the input layer of LSTM.
- **Hierarchical LSTM:** This baseline approach is used for question classification proposed by Xia et al. [14], which uses a hierarchical LSTM model to encode the question texts for classification.

Table 2. *Accuracy* and *Macro-F1* on QA aspect classification.

Approaches	Accuracy	Macro-F1
CNN(A) (Kim et al. [5])	0.575	0.294
CNN(Q)	0.744	0.585
CNN(Q+A)	0.771	0.595
LSTM(A) (Tang et al. [11])	0.675	0.468
LSTM(Q)	0.804	0.665
LSTM(Q+A)	0.850	0.706
Hierarchical LSTM (Xia [14])	0.827	0.729
Individual-Attention (Wang et al. [13])	0.835	0.755
Multi-Attention (ours)	0.865	0.818

- **Individual-Attention:** This baseline approach leverages attention mechanism to capture the relevant information between the question and answer text without clause segmentation proposed by Wang et al. [13].
- **Multi-Attention:** This is our proposed approach which introduces multi-attention mechanism with clause segmentation.

Table 2 demonstrates the experimental results of all approaches in our experiment. By analysis, we can draw some conclusions as follows:

First, by analyzing the approaches based on **CNN** and **LSTM**, we can find that approaches only using question texts as input all outperform those only using answer texts for classification, which accords with annotation guidelines **G1** and **G2** in Sect. 3.

Second, in the approaches **CNN** and **LSTM**, approaches with the concatenation of question and answer texts as input are better than other methods, which demonstrates that answer texts can assist question texts and bring performance improvement for QA aspect classification.

Third, we can find that the performance of approaches based on **LSTM** is obviously superior to that of approaches based on **CNN**. Thus, LSTM is better for QA aspect classification task than CNN.

Therefore, the last three approaches are based on LSTM and take question texts as input. And the two approaches **Individual-Attention** and **Multi-Attention** utilize the relevant information in answer texts based on attention mechanism. The approach **Hierarchical LSTM** performs better than **LSTM** but worse than **Individual-Attention**. And the **Individual-Attention** approach achieves the improvement of 3.1% (Accuracy) and 9.0% (*Macro-F1*) compared with the **LSTM** approach, which proves that it is a good choice to introduce attention mechanism to capture relevant information with respect to the annotated aspect between the question and answer text.

Our proposed **Multi-Attention** approach outperforms most of all approaches, and the accuracy and *Macro-F1* of our model are respectively 3% and 6.3% higher than those of the method **Individual-Attention**. The empirical studies demonstrate that our proposed approach in which we introduce multiple attention representations based on clause segmentation algorithm, can capture the most aspect-related information between the question and answer texts in a QA text pair so as to achieve better performance for QA aspect classification. Significance test shows that this improvement is significant $(p - value < 0.05)$.

5.3   Parameter Analysis

The hyper-parameters tuned according to the development data are optimal and different parameters may affect the performance of our proposed approach. The key parameter in our experiment is the number of clauses in the answer text N_{max}. When the other parameters are fixed, we tune the parameter N_{max} and find that the best value of N_{max} is 3. Figure 3 depicts the line chart of *Accuracy* and *Macro-F* changing with N_{max}.

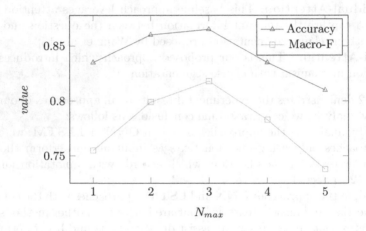

Fig. 3. The line chart of *Accuracy* and *Macro-F* changing with N_{max}.

6   Conclusion

QA aspect classification task is essential to QA aspect-level sentiment analysis. The characteristic of our proposed approach is that we introduce multi-attention mechanism based on clause segmentation to generate multiple attention representations of the question text, which extends the feature representation of the question text to further improve the performance of QA aspect classification. The experimental results demonstrate that the Multi-Attention representation method outperforms some strong baseline methods neural networks.

In our future work, we will consider to perform joint learning with the two tasks QA aspect classification and aspect term extraction to further achieve better performance for QA aspect classification.

Acknowledgements. This work is supported in part by Industrial Prospective Project of Jiangsu Technology Department under Grant No. BE2017081 and the National Natural Science Foundation of China under Grant No. 61572129.

References

1. Chen, Z., Liu, B.: Topic modeling using topics from many domains, lifelong learning and big data. In: International Conference on Machine Learning, pp. 703–711 (2014)
2. Duchi, J., Hazan, E., Singer, Y.: Adaptive subgradient methods for online learning and stochastic optimization. J. Mach. Learn. Res. **12**(Jul), 2121–2159 (2011)
3. Glorot, X., Bengio, Y.: Understanding the difficulty of training deep feedforward neural networks. In: Proceedings of the 13th International Conference on Artificial Intelligence and Statistics, pp. 249–256 (2010)
4. He, R., Lee, W.S., Ng, H.T., Dahlmeier, D.: An unsupervised neural attention model for aspect extraction. In: Proceedings of the 55th Annual Meeting of the Association for Computational Linguistics, vol. 1, pp. 388–397 (2017)
5. Kim, Y.: Convolutional neural networks for sentence classification. In: Proceedings of the 2014 Conference on EMNLP, pp. 1746–1751 (2014)
6. Mitchell, M., Aguilar, J., Wilson, T., Van Durme, B.: Open domain targeted sentiment. In: Proceedings of the 2013 Conference on Empirical Methods in Natural Language Processing, pp. 1643–1654 (2013)
7. Mukherjee, A., Liu, B.: Aspect extraction through semi-supervised modeling. In: Proceedings of the 50th Annual Meeting of ACL, pp. 339–348. ACL (2012)
8. Poria, S., Cambria, E., Ku, L., Gui, C., Gelbukh, A.: A rule-based approach to aspect extraction from product reviews. In: Proceedings of the Second Workshop on Natural Language Processing for Social Media, pp. 28–37 (2014)
9. Rana, T.A., Cheah, Y.: A two-fold rule-based model for aspect extraction. Expert Syst. Appl. **89**, 273–285 (2017)
10. Shu, L., Xu, H., Liu, B.: Lifelong learning CRF for supervised aspect extraction. In: Proceedings of the 55th Annual Meeting of the Association for Computational Linguistics, vol. 2, pp. 148–154 (2017)
11. Tang, D., Qin, B., Feng, X., Liu, T.: Effective LSTMS for target-dependent sentiment classification. In: Proceedings of COLING 2016, the 26th International Conference on Computational Linguistics: Technical Papers, pp. 3298–3307 (2016)
12. Toh, Z., Su, J.: NLANGP: supervised machine learning system for aspect category classification and opinion target extraction. In: International Workshop on Semantic Evaluation, pp. 496–501 (2015)
13. Wang, Y., Huang, M., Zhao, L., et al.: Attention-based LSTM for aspect-level sentiment classification. In: Proceedings of the 2016 Conference on Empirical Methods in Natural Language Processing, pp. 606–615 (2016)
14. Xia, W., Zhu, W., Liao, B., Chen, M., Cai, L., Huang, L.: Novel architecture for long short-term memory used in question classification. Neurocomputing **299**, 20–31 (2018)
15. Xue, W., Zhou, W., Li, T., Wang, Q.: MTNA: a neural multi-task model for aspect category classification and aspect term extraction on restaurant reviews. In: Proceedings of the 8th International Joint Conference on Natural Language Processing, pp. 151–156. Asian Federation of Natural Language Processing (2017)
16. Yang, Y., Liu, X.: A re-examination of text categorization methods. In: International ACM SIGIR Conference on Research and Development in Information Retrieval, pp. 42–49 (1999)

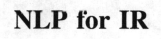

NLP for IR

Hierarchical Answer Selection Framework for Multi-passage Machine Reading Comprehension

Zhaohui Li, Jun Xu[✉], YanYan Lan, Jiafeng Guo, Yue Feng, and Xueqi Cheng

CAS Key Lab of Network Data Science and Technology,
Institute of Computing Technology, Chinese Academy of Sciences, Beijing, China
{lizhaohui,fengyue}@software.ict.ac.cn,
{junxu,lanyanyan,guojiafeng,cxq}@ict.ac.cn

Abstract. Machine reading comprehension (MRC) on real web data, which means finding answers from a set of candidate passages for a question, is a quite arduous task in natural language processing. Most state-of-the-art approaches select answers from all passages or from only one single golden paragraph, which may cause the overlapping information and the lack of key information. To address these problems, this paper proposes a hierarchical answer selection framework that can select **main content** from a set of passages based on the question, and predict final answer within this main content. Specifically, three main parts are employed in this pipeline: First, the passage selection model uses a classification mechanism to select passages by passages content and title information which is not fully used in other models; Second, a key sentences sequence selection mechanism is modeled by Markov-Decision-Process (MDP) in order to gain as much as effectual answer information as possible; Finally, a match-LSTM model is employed to extract the final answer from the selected main content. These three modules that shared the same attention-based semantic network and we conduct experimental on DuReader search dataset. The results show that our framework outperforms the baseline by a large margin.

Keywords: Machine reading comprehension
Markov Decision Process · Reinforcement learning
Natural language process

1 Introduction

In NLP community, Machine reading comprehension (MRC) task which aims to endow the machine with the ability of answering questions after reading a passage or a set of passages has been popular in recent years. At the outset, MRC task dataset was cloze task [2,3], and then was multiple-choice exam [4] which collect multiple-choice questions from exams. Finally, it evolved into answering more complex questions based on single or multiple documents [1,5,8]

© Springer Nature Switzerland AG 2018
S. Zhang et al. (Eds.): CCIR 2018, LNCS 11168, pp. 93–104, 2018.
https://doi.org/10.1007/978-3-030-01012-6_8

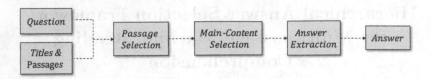

Fig. 1. Overview pipeline of question based hierarchical answer selection framework.

Recently, several significant successes in answer extraction on single passage [6,9,13] help MRC research barging its way into a high level, especially some methods have outperformed human annotators on the SQuAD dataset [8]. These methods formulate MRC task as predicting the start and end positions of the answer in the passage. However, this milestone is not strong enough when considering the real web data. Therefore, some realistic MRC dataset [1,5] are released based on search engines. For every question, the MRC model need to consider all passages related to the question so as to get the result. To handle these complex MRC tasks, most existing methods follow the single passage extraction based approach in the SQuAD dataset, and several novel methods such as v-net [16] and s-net [12] have achieved high performance in MS-MARCO [5] leaderboard.

However, when it comes to DuReader task [1], these methods may not work well for they are ether doing verification between candidate answers or synthesizing candidate answers which generated in a single passage extraction. Compared with MS-MARCO, DuReader is much larger and complex, and each question has multiple candidate passages. DuReader example[1] illustrates that DuReader has more unrelated and overlapping information, some are even incorrect. Moreover, the simple span-prediction in a continuous passage text is unlikely to work well because the components of the answers are relatively far from source passage (not in a single continuous text span). Thus, it is unreasonable to get continuous answer spans from all source passages.

In order to solve this intrinsic challenge for multi-passage MRC, inspired by human MC process, this paper proposes a hierarchical answer selection framework which is designed for MRC on real web data. It can gets **Main Content**, that is the crucial information from the passages according the question' s content, and then the final answer will be selected within this main content by the state-of-the-art single passage MRC method.

The general framework of this model is demonstrated in Fig. 1, which consists of one encoding module and other three function modules. Specifically, the encoding module employs the Bi-Directional Attention mechanism [9] to obtain a question-aware word representation. And then three other modules are implemented. First, we presume that the *is_selected* labeled passage contain all of the answer information, which will be testified in the latter discussion experiment. And the whole passages content is represented by the combination of the title of passage and the sentence sequence in passage. Therefore, a candidate passage selection model can be trained according to *is_selected* label.

---

[1] See examples at https://ai.baidu.com/broad/introduction?dataset=dureader.

Second, a key-sentence selection model is developed using *Markov-Decision-Process*(MDP) [7,11] and *Policy Gradient* [11] to predict a sequence of key sentences, which is called **Main Content** in our approach. Finally, a state-of-the-art model [13] is employed to obtain a answer from the Main Content.

We conduct statistics and experiments on the DuReader [1] datasets. The results show that our hierarchical answer MRC model outperforms the baseline models. According these results, several contributions can be summarized: (1) We formulate the candidate answer selection as a MDP model, which performs well in large search engine dataset. (2) We first use the title of passage to filter the passages, which is a complement for passage representation. (3) We propose a real world MRC task pipeline based on these modules, which can tackle the redundant information problems.

2 Related Work

In resent years, machine reading comprehension has gained more and more attention, and existing main-steam works are building data-drive, end-to-end neural network models.

At first, some datasets for studying machine comprehension were created in Cloze style, and the task is to predict the missing word [2,3]. Then instead of Cloze style, a significant dataset, the SQuAD dataset [8] was also created by human annotators, people have to predict answers from given passages. Mainstream methods are all boundary models [9,13,17], that is treating MRC as extracting answer span from the given passage, which is usually achieved by predicting the start and end position of the answer.

Recently, two multi-passage real world web MRC dataset released: DuReader [1] and MS-MARCO [5]. Some studies [10,15] concatenate those passages and employ the same models designed for single-passage MRC. On the other hand, more and more latest studies start to design special methods for multiple passages. For example, Wang [14] implemented a pipeline method that ranking the passages first and then extract answer from the selected passages. Tan [12] also use the similar method, which treats the passage ranking as an auxiliary task that can be trained jointly with the reading comprehension model. In comparison with these passage selection model, our approach has a unique hierarchical framework, which use the title information to select the candidate passages and then a sentence-level MDP filter is employed to obtain the main content. In addition, there are also some joint training end-to-end models, such as answer verification method v-net [16], that is, extract answers from passages and then do answer verification process. Actually, answer verification method and our model have a same motivation, that is reduce the overlapping information. However, we implemented our framework as three separate step, while they trained their model jointly.

3   Hierarchical Answer Selection Framework

Figure 1 shows the architecture of our hierarchical MRC model which is composed of an attention encoding layer and three selection modules including Passage Selection module, Main Content Selection module and Answer Span Prediction module. The Attention Encoding module uses Bi-Directional Attention mechanism [9] to obtain a question-aware representation for each passage and title. And then is the answer selection model: First, the passage selection module is determined by titles and passages; Second, the main content selection module is modeled by MDP method. Finally, the Answer Span Prediction module which is a state-of-the-art model [13] is employed to obtain a continuous answer span from the Main Content.

3.1   Attention Encoding Module

For every word in data set, its embedding is assigned at first. Then following Seo [9], we calculate the word-level, sentence level and passage level by attention mechanism. Given a question $\mathbf{Q}$, a set of passages $\mathbf{P}$ and the passages' title $\mathbf{T}$, we first map these words with their word-level embeddings. And then three Bi-directional LSTM are employed to get the new contextual representation: $\mathbf{u}_t^Q, \mathbf{u}_t^{P_i}$, and $\mathbf{u}_t^{T_i}$, which represent vectors of the $\mathbf{t}^{th}$ word in $\mathbf{Q}$, $\mathbf{P}_i$, $\mathbf{T}_i$ respectively.

After getting the base representation of each word, one essential step is to endow these passages and titles with their question's information. We conduct these $\mathbf{P}_i - \mathbf{Q}$ and $\mathbf{T}_i - \mathbf{Q}$ Matching with Attention Flow Layer (Seo et al., 2016), so the $\mathbf{T}_i/\mathbf{P}_i$-to-$\mathbf{Q}$ attention can be easily obtained. Then strictly following Seo et al., 2016, we calculate the question-aware passage and title word representations. And $\{\mathbf{P}_i\}$ and $\{\mathbf{T}_i\}$, and $\{\mathbf{Q}\}$ representation and their sentences representations can be easily calculate by $BiLSTM$ mechanism with the word representations.

3.2   Passage Selection Module

In this part, we formulate the passage representation as the combination of passage title and passage context. So the real representation of each passage is:

$$\tilde{P}_i = [T_i; P_i] \tag{1}$$

where [;] is vector concatenation across row, T_i is the representation of passage title and P_i is the representation of passage content. After that, with these results $\tilde{P}_i$ and Q, we use a simplified classification model to calculate the probability p_i of whether a passage is *is_selected* labeled:

$$p_i = \sigma(Q^T W \tilde{P}_i + b) \tag{2}$$

where $\mathbf{W} \in \mathbb{R}^{d \times d}$ and $b \in \mathbb{R}$, and we use the question-aware passage representation $\tilde{P}_i$ and Q. Given the set of N training samples, each question contains $|P|$

passages with ground truth of *is_selected* label: $(y_1, y_2, \ldots, y_N)$, it can be trained by minimizing negative as the averaged cross entropy loss:

$$\mathcal{L}_s = -\frac{1}{N}\frac{1}{|P|}\sum_{j=1}^{N}\sum_{i=1}^{|P|}[y_i \log p_i + (1 - y_i)(1 - \log p_i)] \tag{3}$$

3.3 Main Content Selection Module

After getting the candidate passages for each question, we employ Markov Decision Process(MDP) to model answer candidate sentence selection. The difficulties lie in how to formalize MRC under the MDP framework. In addition, how to convert the reference answer to the supervision information that can be utilized by MDP model is also a tough question.

MDP Formulation of Sentence Selection. In the encoding module, we have got the representation of each passage $\mathbf{P}$ and its sub-sentence list $X = \{\mathbf{x}_1, \cdots, \mathbf{x}_M\} \subseteq \mathcal{X}$, and $\mathcal{X}$ is the set of all sentences. The goal of sentence selection is to construct a model which can give a set of candidate sentences set so that the following model can find the best answer from this sentence set. The training of a sentence selection model, thus, can be consider as the learning parameters in a MDP model, in which each step corresponds to a selected candidate answer sentence. The states, actions, rewards, transitions, and policy of MDP are set as:

States $\mathcal{S}$: State are designed at step t as a triple $s_t = [\mathbf{Q}, \mathcal{Z}_t, X_t]$, where $\mathbf{Q}$ is the preliminary representation of the question; $\mathcal{Z}_t = \{\mathbf{x}_{(n)}\}_{n=1}^{t}$ is the sequence of t preceding sentences, where $\mathbf{x}_{(n)}$ is the tth sentence in the main content sequence; X_t is the set of candidate sentences. At the beginning $(t = 0)$, the state is initialized as $s_0 = [\mathbf{q}, \emptyset, X]$, where $\emptyset$ is the empty sequence and X contains all of the M sentences in all the candidate passages. Note that we require that each sentence set ends with a special end-of-content symbol $\langle EOS \rangle$, which enables the model to define a distribution over sequences of all possible lengths.

Actions $\mathcal{A}$: At each time step t, the $\mathcal{A}(s_t)$ is the set of actions the agent can choose, each corresponds to a sentence from X_t. That is, the action $a_t \in \mathcal{A}(s_t)$ at the time step t selects a sentence $\mathbf{x}_{m(a_t)} \in X_t$ for the main content sequence, where $m(a_t)$ is the index of the sentence selected by a_t.

Transition T: The transition function $T : \mathcal{S} \times \mathcal{A} \rightarrow \mathcal{S}$ is defined as follows:

$$\begin{aligned}
s_{t+1} = T(s_t, a_t) &= T([\mathbf{Q}, \mathcal{Z}_t, X_t], a_t) \\
&= [\mathbf{Q}, \mathcal{Z}_t \oplus \{\mathbf{x}_{m(a_t)}\}, X_t \setminus \{\mathbf{x}_n\}_o^{a_t}],
\end{aligned} \tag{4}$$

where $\oplus$ appends $\mathbf{x}_{m(a_t)}$ to $\mathcal{Z}_t$ and $\setminus$ removes $\mathbf{x}_n\}_o^{a_t}$ from X_t, o is the first sentence number in X_t. At each time step t, based on state s_t the system chooses an action a_t. Then, the system moves to step $t + 1$ and the system transits to a

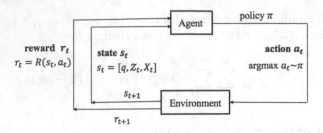

Fig. 2. The agent-environment model in MDP.

new state s_{t+1}: The selected sentence is appended to the end of $\mathcal{Z}_t$, generating a new sentence sequence, and the sentences in the precede place of the selected sentence at step t are removed from the candidate set: $X_{t+1} = X_t \setminus \{x_n\}_0^{a_t}$ (Fig. 2).

Reward R: The reward can be considered as the evaluation of the information quality of the main content sequence, for we aimed at maximize the information related to the answer in this main content selection module. Since we need to maximize the information, we define the reward function on the combination of F1 and Rouge-L:

$$\mathbf{r}(t) = \frac{1}{2} * [\text{F1}(t) + \text{Rouge-L}(t)] \tag{5}$$

where t is the tth sentence in the main content, and the position 0 is defined as zero. Then the reward function caused by choosing the action a_t is:

$$\mathbf{R}(s_t, a_t) = \mathbf{r}(t+1) - \mathbf{r}(t) \tag{6}$$

Policy function p: The policy $\mathbf{p}(s)$ defines a function that takes the state as input and output a distribution over all of the possible actions $a \in \mathcal{A}(s)$. Specifically, each probability in the distribution is a normalized function whose input is the bilinear product of the LSTM function and the selected sentence:

$$p(a|s) = \frac{\exp\left\{\mathbf{x}_{m(a)}^T \mathbf{U}_p \, \text{LSTM}(s)\right\}}{\sum_{a' \in \mathcal{A}(s)} \exp\left\{\mathbf{x}_{m(a')}^T \mathbf{U}_p \, \text{LSTM}(s)\right\}} \tag{7}$$

where The deep neural network model $LSTM : \mathcal{S} \rightarrow \mathbb{R}^L$ maps a state to a real vector where L is the number of dimensions. Given $s = [\mathbf{Q}, \mathcal{Z} = \{\mathbf{x}_1, \mathbf{x}_2, \cdots, \mathbf{x}_t\}, X_t]$, where $\mathbf{x}_k(k = 1, \cdots, t)$ is the sentence at k-th position and represented with its embedding. Thus, the policy function $\mathbf{p}(s)$ is:

$$\mathbf{p}(s) = \langle p(a_1|s), \cdots, p(a_{|\mathcal{A}(s)|}|s) \rangle. \tag{8}$$

Learning with Policy Gradient. The model has some parameters to learn, we donate as Θ. In this training phase, suppose there are N training questions

---

Algorithm 1. MDP-MCS

Input: Training set $D = \{(Q^{(n)}, X^{(n)}, A^{(n)})\}_{n=1}^{N}$, learning rate η, dropout keep rate d, and value
    function R
Output: Θ
1: Initialize $\Theta \leftarrow$ random values in $[-1, 1]$
2: **repeat**
3:    **for all** $(Q, X, A) \in D$ **do**
4:       $(s_0, a_0, r_1, \cdots, s_{M-1}, a_{M-1}, r_M) \leftarrow SampleEpisode(\Theta, Q, X, A, R) \{Algorithm(2)\}$
5:       **for** $t = 0$ to $M - 1$ **do**
6:          $G_t \leftarrow \sum_{k=0}^{M-1-t} r_{t+k+1} \{Equation(17)\}$
7:          $\Theta \leftarrow \Theta - \eta G_t \nabla_\Theta \log a_t | s_t; \Theta \{\text{According to Equation}(18)\}$
8:       **end for**
9:    **end for**
10: **until** converge
11: **return** Θ

$\{(Q^{(n)}, X^{(n)}, A^{(n)})\}_{n=1}^{N}$, where $A^{(n)}$ denotes the reference answers of the question. Inspired by the RL algorithm policy gradient, we devised a novel algorithm which can learn the parameters toward the Main Content selection Model. It is referred as MDP-MCS and shown in Algorithm 1. The Algorithm 2 shows the procedure of sampling a sentence episode for Algorithm 1. The definition of long-term return G_t is crucial important, for it equals the ground truth in this task. So we define the discounted sum of rewards from position t as G_t:

$$G_t = \sum_{k=0}^{M-1-t} \lambda r_{t+k+1} \tag{9}$$

where M is the length of the selected sentence episode and λ is the discount rate of policy gradient. Note that in our model, $\lambda = 1$. And using the long-term return G_t, we can calculate the loss of each iteration, an sentence episode(consisting a sequence of states, actions, and rewards) is sampled according to current policy.

$$\mathcal{L}(E) = -\sum_{t=1}^{|E|} \left(\sum_{a \in \mathcal{A}(s_t)} G_t(a|s_t) \log \frac{1}{p(a|s_t)} \right). \tag{10}$$

Testing Method. After the training phase, we can get an agent that can select main content sentences from passages according to its policy function. Specifically, given a question Q, a set of M candidate sentence X, the system state is initialized as $s_0 = [Q, \mathcal{Z}_0 = \emptyset, X_0 = X]$. Then, at each of the steps $t = 0, \cdots, M - 1$, the agent receives the state $s_t = [Q, \mathcal{Z}_t, X_t]$ and searches the policy π, on the basis of policy function $\mathbf{p}$. Then, it chooses an action a according to π. Moving to the next step $t + 1$, the state becomes $s_{t+1} = [q, \mathcal{Z}_{t+1}, X_{t+1}]$. The process is repeated until the candidate set becomes empty.

    This MDP-MCS Reinforcement learning method can imitate the reading process of human being. It formulates the main content selection process as a sequence selection episode step by step. By this method, answer informations

---

Algorithm 2. Sample Episode

Input: Parameters Θ, question Q, candidate passage sentences X, reference answers
    A, and value function R
Output: A selected sentence Episode
 1: Initialize $\Theta \leftarrow$ random values in $[-1, 1]$
 2: $s \leftarrow [\mathbf{Q}, \emptyset, X]$
 3: $E = ()\{\text{empty episode}\}$
 4: **while** $\mathbf{x}_{m(a)}$ is not $< EOS >$ **do**
 5:    $\mathcal{A} \leftarrow \mathcal{A}(s)\{\text{Possible actions according to X in state s}\}$
 6:    **for all** $a \in \mathcal{A}$ **do**
 7:        $P(a) \leftarrow \pi(a|s; \Theta)$
 8:    **end for**
 9:    $\tilde{a} = \arg\max_{a \in A(s)} \pi(a|s)$ {Sample an action $\tilde{a} \in \mathcal{A}$ according to P}
10:    $r \leftarrow \mathbf{R}(s, \tilde{a})\{\text{Calculate on the basis of } A\}$
11:    $E \leftarrow E \oplus \{(s, \tilde{a}, r)\}$
12:    $s \leftarrow [s, \mathbf{Q}, \mathcal{Z} \oplus \{\mathbf{x}_{m(a)}\}, X \setminus \{\mathbf{x}_n\}_0^{m(a)}]$
13: **end while**
14: **return** $E = (s_0, a_0, r_1, \cdots, s_{M-1}, a_{M-1}, r_M)$

---

which have not in the same paragraph can be selected and the overlapping information can be filtered. These are the merits of MDP-MCS model.

3.4 Answer Span Prediction Module

In order to extract the final answer from the selected main content, we employed a main-steam boundary model to local the answer span. Pointer Network and Match-LSTM are used to compute the probability of each word:

$$\mathbf{g}_t^k = w_1 \tanh\left(w_2 \left[\mathbf{v}_t^C, \mathbf{h}_{k-1}\right]\right) \qquad \mathbf{h}_k = \mathbf{LSTM}\left(h_{k-1}, c_k\right) \qquad (11)$$

$$\alpha_t^k = \exp\left(g_t^k\right) \sum_{j=1}^{|C|} \exp\left(g_t^k\right) \qquad \mathbf{c}_k = \sum_{t=1}^{|C|} \alpha_t^k \mathbf{v}_t^C \qquad (12)$$

where α_t^1 and α_t^2 is the probability of the tth word in the passage to be the start and the end position of the answer span. C is the main content selected by the last sub-section v_t^C is the new main content word representation calculated by the first encode module. This boundary model can be trained by minimizing the negative log probabilities of the true start and end positions:

$$\mathcal{L}_{boundary} = -\frac{1}{N} \sum_{i=1}^{N} \left(\log \alpha_{y_i^1}^1 + \log \alpha_{y_i^2}^2\right). \qquad (13)$$

where y_i^1 and y_i^2 are the gold start and end positions, N is the scale of dataset.

4 Experiment

4.1 Dataset and Evaluation Metrics

Considering the large scale of DuReader dataset (The training, development and test sets consist of 181K, 10K and 10K questions, 855K, 45K and 46K documents, 376K, 20K and 21K answers, respectively.) and the time consumption of policy gradient method, we use about half of the train dataset which is classified as 'Baidu Search' dataset by DuReader. And we evaluate the reading comprehension task via BLEU-4 and Rouge-L, which are widely used for evaluating the quality of language generation. And for the main content selection model, $F1$ and *Recall* are also used as evaluation method.

4.2 Implementation Details

First, we pre-train the corpus with Glove[2] as the initial embeddings, and words whose count number is less than 5 will not be involved in the vocabulary. Our models are optimized using Adam algorithm with a initial learning rate as 0.001 and dropout rate 0.6. And in the passage selection phase, we simply treat the question with no selected passage or empty main content as the *No-Answer* question. In the main content selection layer, we use zero vector to represent the end sentence $<EOS>$. Besides, the word embedding size is 300-dimension and all hidden state sizes is 150-dimension. Note that we do not initialize the model parameters every times, the last train parameters are used in the next train time. When training the finally answer extraction module, we choose the text span in the main content with the highest BLEU-4 score as gold span. Moreover, we only use the main content whose ROUGE-L score is higher than 0.7.

4.3 Experimental Results

We compared our method with several state-of-the-arts baselines in MRC. The results is demonstrated in Table 1: The first part is the single passage selection methods; The second part are the boosting models and the multi-passage MRC method including our model.

GP: Golden Paragraph, a heuristic approach which chooses paragraph has the largest overlap with the answers in a document as answers. In testing phase, choosing paragraph which have the largest overlap with the question as answer. **Match-LSTM** [13]: Using Pointer Network and Match-LSTM to predict the beginning or ending points in the whole passages set. **BiDAF** [9]: a method which employed a bi-directional attention flow mechanism to achieve a question-aware context representations for the passage, then the beginning and ending points were predict based on the representations. **PR+BiDAF**: Using Passage Ranking to select the passages and then using BiDAF model to predict the

---

[2] Pre-trained word vectors (http://nlp.stanford.edu/data/glove.6B.zip).

answer. **V-Net** [16]: a method which extract candidate answers from all passages, and then do verification among those candidate answers. **S-Net** [12]: a model that consists of evidence extraction part and answer synthesis part.

The results of several baseline systems and our model are shown in Table 1. We can see that the *GP* method can improve the baseline methods significantly, but it cannot beat our main content selection model for there is no reference answer in the test dataset and simply matching the question words and with passage paragraph cannot lead to a better performance. The passage ranking method cannot outperform ours as well, for we get the selected sentences information within the selected passages.

Table 1. Performance of all methods on DuReader search test dataset.

Method	BLEU-4%	ROUGE-L%
GP	27.7	60.2
Match-LSTM	23.1	31.2
BiDAF	23.1	31.1
PR+BiDAF	37.55	41.81
V-Net	40.97	44.18
S-Net	41.12	44.52
GP+Match-LSTM	39.99	44.15
GP+BiDAF	39.83	42.01
Our model	**42.68**	**44.95**

5   Analysis and Discussion

In this section, we conduct experiments to show the reasons why our hierarchical answer selection model outperformed the baselines. Since answers on the test set are not published, we analyze our model on the development set.

5.1   Reasoning of Passage Selection Module

In theory, there are two reasons why the passage selection module and main content selection module are effective on DuReader Search dataset. First, we do not need read all the passages to get our answers when reading especially in test process. Second, some real-world answer passages contain the wrong answers or irrelevant answers which are noise in the answer span predict model. DuReader train and dev datasets are the real-world data collected by Baidu Search Engine, so each passages in them has the *is_selected* label which can provide ground truth for passage selection, we conduct statistics experiments to prove it. From the example in the website[3] we can see that the candidate passages are always about

---

[3] DuReader dataset(https://ai.baidu.com/broad/introduction?dataset=dureader).

Table 2. Statistics of the relation between selected passages and answers of question

Dataset	Answer in selected passage	Answer not in selected passage	Total
Train	87502	3706	91208
Dev	4883	117	5000

the same theme and some even talk about unrelated things. And by mathematical statistics (Table 2) of train and dev dataset, we found that the *is_selected* passages almost contain all of the answer information. And Table 3 demonstrated that the passage selection layer do improve the performance of our model, for trash and overlapping information will influence the following sentence selection MDP model and the answer span predict model.

Table 3. Comparison of each module of our model on DuReader search dev dataset.

Method	F1%	Recall%	ROUGE-L%	BLEU-4%
Passage selection	9.05	90.44	25.56	9.90
MDP-MCS	10.14	60.21	21.30	10.63
PS+MDP-MCS	22.14	86.42	**62.5**	**30.50**

5.2 Necessity of Main Content Selection Module

After getting the selected passages, we use MDP-MCS model to selected sentence-level answer information, for real-world dataset answer information may not in a continuous span of a passage, and answers always have a far edit distance. That is why some methods which work well in Ms-Marco do not performance well in DuReader dataset. And the experiment in Table 3 shows MDP-MCS combined with PS model can significantly improve the F1 and Recall score, which means it can select main content effectively. In addition, the use of MDP-MCS based on passage selection could significantly boosts the overall performance(ROUGE-L and BLEU-4) of answer span predict model. But the single MDP-MCS does not work well because the MDP model cannot handle a long sequence, which means that passages must be filtered before this layer. So it is necessary to implement a passage selection model and a main content selection model in a MRC pipeline.

6   Conclusion

In this paper, we propose a hierarchical answer selection framework pipeline to tackle the multi-passage MRC task. This framework contains three modules: The first module, passage selection layer first use the passage title and content information to predicted which passage will be selected; Then a main content selection module is modeled by MDP and trained by policy gradient; Finally, using Match-LSTM method, final answer can be generated. This hierarchical answer selection framework has achieved the state-of-the-art performance on a challenging dataset DuReader, which is designed for MRC on real web data.

References

1. He, W., et al.: DuReader: a Chinese machine reading comprehension dataset from real-world applications. arXiv preprint arXiv:1711.05073 (2017)
2. Hermann, K.M., et al.: Teaching machines to read and comprehend. In: Advances in Neural Information Processing Systems, pp. 1693–1701 (2015)
3. Hill, F., Bordes, A., Chopra, S., Weston, J.: The goldilocks principle: reading children's books with explicit memory representations. arXiv preprint arXiv:1511.02301 (2015)
4. Lai, G., Xie, Q., Liu, H., Yang, Y., Hovy, E.: Race: large-scale reading comprehension dataset from examinations. arXiv preprint arXiv:1704.04683 (2017)
5. Nguyen, T., et al.: MS MARCO: a human generated machine reading comprehension dataset. arXiv preprint arXiv:1611.09268 (2016)
6. Pan, B., Li, H., Zhao, Z., Cao, B., Cai, D., He, X.: MEMEN: multi-layer embedding with memory networks for machine comprehension. arXiv preprint arXiv:1707.09098 (2017)
7. Puterman, M.L.: Markov Decision Processes: Discrete Stochastic Dynamic Programming. Wiley, Hoboken (2014)
8. Rajpurkar, P., Zhang, J., Lopyrev, K., Liang, P.: SQuAd: 100,000+ questions for machine comprehension of text. arXiv preprint arXiv:1606.05250 (2016)
9. Seo, M., Kembhavi, A., Farhadi, A., Hajishirzi, H.: Bidirectional attention flow for machine comprehension. arXiv preprint arXiv:1611.01603 (2016)
10. Shen, Y., Huang, P.S., Gao, J., Chen, W.: ReasoNet: learning to stop reading in machine comprehension. In: Proceedings of the 23rd ACM SIGKDD International Conference on Knowledge Discovery and Data Mining, pp. 1047–1055. ACM (2017)
11. Sutton, R.S., Barto, A.G.: Reinforcement Learning: An Introduction, vol. 1. MIT Press, Cambridge (1998)
12. Tan, C., Wei, F., Yang, N., Lv, W., Zhou, M.: S-Net: from answer extraction to answer generation for machine reading comprehension. arXiv preprint arXiv:1706.04815 (2017)
13. Wang, S., Jiang, J.: Machine comprehension using match-LSTM and answer pointer. arXiv preprint arXiv:1608.07905 (2016)
14. Wang, S., et al.: Reinforced reader-ranker for open-domain question answering. arXiv preprint arXiv:1709.00023 (2017)
15. Wang, S., et al.: Evidence aggregation for answer re-ranking in open-domain question answering. arXiv preprint arXiv:1711.05116 (2017)
16. Wang, Y., et al.: Multi-passage machine reading comprehension with cross-passage answer verification. arXiv preprint arXiv:1805.02220 (2018)
17. Xiong, C., Zhong, V., Socher, R.: Dynamic coattention networks for question answering. arXiv preprint arXiv:1611.01604 (2016)

Generative Paragraph Vector

Ruqing Zhang[1,2](✉), Jiafeng Guo[1,2], Yanyan Lan[1,2], Jun Xu[1,2],
and Xueqi Cheng[1,2]

[1] University of Chinese Academy of Sciences, Beijing, China
[2] CAS Key Lab of Network Data Science and Technology,
Institute of Computing Technology, Chinese Academy of Sciences, Beijing, China
zhangruqing@software.ict.ac.cn,
{guojiafeng,lanyanyan,junxu,cxq}@ict.ac.cn

Abstract. The recently introduced Paragraph Vector (PV) is an efficient method for learning high-quality distributed representations for texts. However, from the probabilistic view, PV is not a complete model since it only models the generation of words but not texts, leading to two major limitations. Firstly, without a text-level model, PV assumes the independence between texts and thus cannot leverage the corpus-wide information to help text representation learning. Secondly, without the generation model of texts, the inference of text representations outside of the training set becomes difficult. Although PV makes itself as an optimization problem so that one can obtain representations for new texts anyway, it loses the sound probabilistic interpretability in that way. To tackle these problems, we first introduce a Generative Paragraph Vector, an extension of the Distributed Bag of Words version of Paragraph Vector with a complete generative process. By defining the generation model over texts, we further incorporate text labels into the model and turn it into a supervised version, namely Supervised Generative Paragraph Vector. Experiments on five text classification benchmark collections show that both unsupervised and supervised model architectures can yield superior classification performance against the state-of-the-art counterparts.

1 Introduction

A central problem in many text based applications, *e.g.*, sentiment classification [16], question answering [22] and machine translation [20], is how to capture the essential meaning of a piece of text in a fixed-length vector. Perhaps the most popular fixed-length vector representations for texts is the bag-of-words (or bag-of-n-grams) [3]. Besides, probabilistic latent semantic indexing (PLSI) [5] and latent Dirichlet allocation (LDA) [1] are two widely adopted alternatives.

A recent paradigm in this direction is to use a distributed representation for texts [14,15]. In particular, [10] show that their method, Paragraph Vector (PV), can capture text semantics in dense vectors and outperform many existing representation models. In PV, the paragraph vector is learned through a word generation/prediction task, *i.e.*, to maximize the probability of words given the

© Springer Nature Switzerland AG 2018
S. Zhang et al. (Eds.): CCIR 2018, LNCS 11168, pp. 105–118, 2018.
https://doi.org/10.1007/978-3-030-01012-6_9

text (and the context words). However, from the probabilistic generation view, PV is not a complete model since it only models the generation of words but not texts, leading to two major limitations: (1) Texts from the same corpus are assumed to be independent from each other and no corpus-wide information/constraint is leveraged in text modeling. (2) Without the generation model of texts, PV suffers a similar problem as PLSI that it is unclear how to infer the text representations outside of the training set with the learned model. PV circumvents this by making itself as an optimization problem so that one can obtain representations for new texts anyway, but this procedure loses sound probabilistic interpretability.

Inspired by the completion and improvement of LDA over PLSI, we first introduce the Generative Paragraph Vector (GPV) with a complete generation process for a corpus. Specifically, GPV can be viewed as an extension of the Distributed Bag of Words version of Paragraph Vector (PV-DBOW), where the paragraph vector is viewed as a hidden variable sampled from some prior distributions, and the words within the text are then sampled from the softmax distribution given the text and word vectors. With a complete generative process, we are able to employ the corpus-wide constraint (*i.e.*, the prior distribution over paragraph vectors) to help regularize the text representation. Meanwhile, by modeling the generation of the paragraph vectors, we are able to infer the representations of new texts based on the learned model.

More importantly, with the ability to infer the distributed representations for unseen texts, we now can directly incorporate labels paired with the texts into the model to guide the representation learning, and turn the model into a supervised version, namely Supervised Generative Paragraph Vector (SGPV). By learning the SGPV model, we can directly employ SGPV to predict labels for new texts. As we know, when the goal is prediction, fitting a supervised model would be a better choice than learning a general purpose representations of texts in an unsupervised way. We further show that SGPV can be easily extended to accommodate n-grams so that we can take into account word order information, which is important in learning semantics of texts.

We evaluated our proposed models on five text classification benchmark datasets. For the unsupervised GPV, we show that its superiority over the existing counterparts, such as bag-of-words, LDA, PV and FastSent [4]. For the SGPV model, we take into comparison both traditional supervised representation models, *e.g.*, MNB [23], and a variety of state-of-the-art deep neural models for text classification [6,8,9,19]. Again we show that the proposed SGPV can outperform the baseline methods by a substantial margin, demonstrating it is a simple yet effective model.

The rest of the paper is organized as follows. We first review the related work in Sect. 2 and briefly describe PV in Sect. 3. We then introduce the unsupervised generative model GPV and supervised generative model SGPV in Sect. 4 respectively. Experimental results are shown in Sect. 5 and conclusions are made in Sect. 6.

2 Related Work

Many text based applications require the text input to be represented as a fixed-length feature vector. The most common fixed-length representation is bag-of-words (BoW) [3]. In the popular TF-IDF scheme, each document is represented by *tfidf* values of a set of selected feature-words. However, the BoW representation often suffers from data sparsity and high dimension. Meanwhile, due to the independent assumption between words, BoW representation has very little sense about the semantics of the words.

To address this shortcoming, several methods are proposed by projecting the texts into a latent topical space, including matrix factorization methods such as Latent Semantic Indexing (LSI) [2], and probabilistic topical models such as Probabilistic Latent Semantic Indexing (PLSI) [5] and Latent Dirichlet Allocation (LDA) [1]. Both PLSI and LDA have a good statistical foundation and proper generative model of the documents, as compared with LSI which relies on a singular value decomposition over the term-document co-occurrence matrix. In PLSI, each word is generated from a single topic, and different words in a document may be generated from different topics. While PLSI makes great effect on probabilistic modeling of documents, it is not clear how to assign probability to a document outside of the training set with the learned model. To address this issue, LDA is proposed by introducing a complete generative process over the documents, and demonstrated as a state-of-the-art document representation method. To further tackle the prediction task, Supervised LDA [13] is developed by jointly modeling the documents and the labels. However, these traditional topic methods are built upon the bag-of-words (BoW) representation, which limits their ability to leverage the rich semantic relatedness of the words.

Recently, distributed models have been demonstrated as efficient methods to acquire semantic representations of texts. A representative method is Word2Vec [14], which can learn meaningful word representations in an unsupervised way from large scale corpus. To represent sentences or documents, a simple approach is then using a weighted average of all the words. A more sophisticated approach is combing the word vectors in an order given by a parse tree [18]. Later, Paragraph Vector (PV) [10] is introduced to directly learn the distributed representations of sentences and documents. There are two variants in PV, namely the Distributed Memory Model of Paragraph Vector (PV-DM) and the Distributed Bag of Words version of Paragraph Vector (PV-DBOW), based on two different model architectures. Although PV is a simple yet effective distributed model on sentences and documents, it only models the generation of words but not texts.

Besides these unsupervised representation learning methods, there have been many supervised deep models with directly learning sentence or document representations for the prediction tasks. Recursive Neural Network (RecursiveNN) [19] has been proven to be efficient in terms of constructing sentence representations. Recurrent Neural Network (RNN) [21] can be viewed as an extremely deep neural network with weight sharing across time. Convolution Neural Network (CNN) [9] can fairly determine discriminative phrases in a text with a max-pooling layer. However, these deep models are usually quite complex and thus the training would be time-consuming on large corpus.

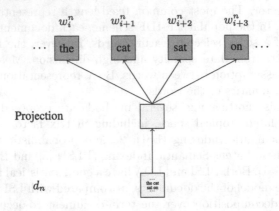

Fig. 1. Distributed Bag of Words version of paragraph vectors. The paragraph vector is used to predict the words in a small window ("the", "cat", "sat" and "on").

3   Paragraph Vector

Since our model can be viewed as a probabilistic extension of the PV-DBOW model with a complete probabilistic foundation, we first briefly review the PV-DBOW model for reference.

In PV-DBOW, each text is mapped to a unique paragraph vector and each word is mapped to a unique word vector in a continuous space. The paragraph vector is used to predict target words randomly sampled from the paragraph as shown in Fig. 1. More formally, Let $D = \{d_1, \ldots, d_N\}$ denote a corpus of N texts, where each text $d_n = (w_1^n, w_2^n, \ldots, w_{l_n}^n), n \in 1, 2, \ldots, N$ is an l_n-length word sequence over the word vocabulary V of size M. Each text $d \in D$ and each word $w \in V$ is associated with a vector $d \in \mathbb{R}^K$ and $w \in \mathbb{R}^K$, respectively, where K is the embedding dimensionality. The predictive objective of the PV-DBOW for each word $w_i^n \in d_n, i = 1, 2, \ldots, l_n$ is defined by the softmax function

$$p(w_i^n|d_n) = \frac{\exp(w_i^n \cdot d_n)}{\sum_{w' \in V} \exp(w' \cdot d_n)}. \tag{1}$$

The PV-DBOW model can be efficiently trained using the stochastic gradient descent (SGD) with negative sampling [15], and the gradient is obtained via backpropagation.

As compared with traditional topic models, e.g. PLSI and LDA, PV-DBOW conveys the following merits. Firstly, PV-DBOW using negative sampling can be interpreted as a matrix factorization over the words-by-texts co-occurrence matrix with shifted-PMI values [11]. In this way, more discriminative information (i.e., PMI) can be modeled in PV as compared with the generative topic

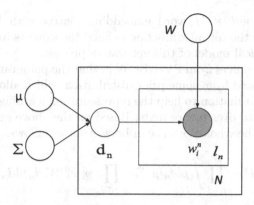

Fig. 2. A graphical model representation of Generative Paragraph Vector (GPV). (The boxes are "plates" representing replicates; a shaded node is an observed variable; an unshaded node is a hidden variable.)

models which learn over the words-by-texts co-occurrence matrix with raw frequency values. Secondly, PV-DBOW does not have the explicit "topic" layer and allows words automatically clustered according to their co-occurrence patterns during the learning process. In this way, PV-DBOW can potentially learn much finer topics than traditional topic models given the same hidden dimensionality of texts. However, without a text-level model, PV-DBOW cannot leverage the corpus-wide information to help text representation learning and make itself heuristic in inferring representations for new texts.

4   Our Models

4.1   Generative Paragraph Vector

In this section, we introduce the Generative Paragraph Vector model in detail. Overall, GPV is a generative probabilistic model for a corpus. We assume that for each text, a latent paragraph vector is first sampled from some prior distributions, and the words within the text are then generated from the normalized exponential (*i.e.* softmax) distribution given the paragraph vector and word vectors. In our work, multivariate normal distribution is employed as the prior distribution for paragraph vectors. It could be replaced by other prior distributions and we will leave this as our future work. The specific generative process is as follows:

For each text $d_n \in D, n = 1, 2, \ldots, N$:
(a) Draw paragraph vector $\boldsymbol{d_n} \sim \mathcal{N}(\mu, \Sigma)$
(b) For each word $w_i^n \in d_n, i = 1, 2, \ldots, l_n$:
Draw word vector $\boldsymbol{w_i^n} \sim \text{softmax}(\boldsymbol{d_n} \cdot W)_i$

where W denotes a $k \times M$ word embedding matrix with $W_{*j} = \boldsymbol{w}_j$, and softmax$(\boldsymbol{d}_n \cdot W)_i$ is the softmax function defined the same as in Eq. (1). Figure 2 provides the graphical model of this generative process.

Note that GPV differs from PV-DBOW in that the paragraph vector is a hidden variable generated from some prior distribution, which allows us to leverage the corpus-wide information to help the representation learning and also to infer the paragraph vector over future texts. Based on the above generative process, the probability of the whole corpus can be written as follows:

$$p(D) = \prod_{n=1}^{N} \int p(d_n | \mu, \Sigma) \prod_{w_i^n \in d_n} p(w_i^n | W, d_n) dd_n. \tag{2}$$

To learn the model, direct maximum likelihood estimation is not tractable due to non-closed form of the integral. We approximate this learning problem by using MAP estimates for $\boldsymbol{d}_n$, which can be formulated as follows:

$$(\mu^*, \Sigma^*, W^*) = \arg \max_{\mu, \Sigma, W} \prod p(\hat{d}_n | \mu, \Sigma) \prod_{w_i^n \in d_n} p(w_i^n | W, \hat{d}_n), \tag{3}$$

where $\hat{d}_n$ denotes the MAP estimate of $\boldsymbol{d}_n$ for d_n, and (μ^*, Σ^*, W^*) denotes the optimal solution. Note that for computational simplicity, in this work we fixed μ as a zero vector and Σ as a identity matrix. In this way, all the free parameters to be learned in our model are word embedding matrix W. By taking the logarithm and applying the negative sampling idea to approximate the softmax function, we obtain the final learning problem

$$\mathcal{L} = \sum_{n=1}^{N} \left(-\frac{1}{2} \|\hat{d}_n\|^2 + \sum_{w_i^n \in d_n} \left(\log \sigma(\boldsymbol{w}_i^n \cdot \hat{d}_n) + k \cdot \mathbb{E}_{w' \sim P_{\mathrm{nw}}} \log \sigma(-\boldsymbol{w}' \cdot \hat{d}_n) \right) \right), \tag{4}$$

where $\sigma(x) = 1/(1+\exp(-x))$, k is the number of "negative" samples, w' denotes the sampled word and P_{nw} denotes the distribution of negative word samples. As we can see from the final objective function, the prior distribution over paragraph vectors actually act as a regularization term. From the view of optimization, such regularization term could constrain the learning space using the corpus-wide information and usually produces better paragraph vectors.

For optimization, we use coordinate ascent, which first optimizes the word vectors W while leaving the MAP estimates $(\hat{d})$ fixed. Then we find the new MAP estimate for each document while leaving the word vectors fixed, and continue this process until convergence. To accelerate the learning, we adopt a similar stochastic learning framework as in PV which iteratively updates W and estimates $\boldsymbol{d}$ by randomly sampling text and word pairs.

At prediction time, given a new text, we perform an inference step to compute the paragraph vector for the input text. In this step, we freeze the vector representations of each word, and apply the same MAP estimation process of $\boldsymbol{d}$ as in the learning phase. With the inferred paragraph vector of the test text, we can feed it to other prediction models for different applications.

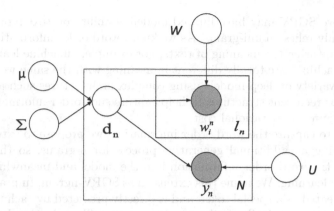

Fig. 3. Graphical model representation of Supervised Generative Paragraph Vector (SGPV).

4.2 Supervised Generative Paragraph Vector

With the ability to infer the distributed representations for unseen texts, we now can incorporate the labels paired with the texts into the model to guide the representation learning, and turn the model into a more powerful supervised version directly towards prediction tasks. Specifically, we introduce an additional label generation process into GPV to accommodate text labels, and obtain the Supervised Generative Paragraph Vector (SGPV) model. Formally, in SGPV, the n-th text $\boldsymbol{d}_n$ and the corresponding class label $y_n \in \{1, 2, \ldots, C\}$ arise from the following generative process:

> For each text $d_n \in \boldsymbol{D}, n = 1, 2, \ldots, N$:
> (a) Draw paragraph vector $\boldsymbol{d}_n \sim \mathcal{N}(\mu, \Sigma)$
> (b) For each word $w_i^n \in d_n, i = 1, 2, \ldots, l_n$:
>   Draw word vector $\boldsymbol{w}_i^n \sim \text{softmax}(\boldsymbol{d}_n \cdot W)_i$
> (c) Draw label $y_n | d_n, U, b \sim \text{softmax}(U \cdot \boldsymbol{d}_n + b)$

where U is a $C \times K$ matrix for a dataset with C output labels, and b is a bias term.

The graphical model of the above generative process is depicted in Fig. 3. SGPV defines the probability of the whole corpus as follows:

$$p(D) = \prod_{n=1}^{N} \int p(d_n | \mu, \Sigma) \Big(\prod_{w_i^n \in d_n} p(w_i^n | W, d_n) \Big) p(y_n | d_n, U, b) d d_n. \tag{5}$$

We adopt a similar learning process as GPV to estimate the model parameters. Since the SGPV includes the complete generative process of both texts and labels, we can directly leverage it to predict the labels of new texts. Specifically, at prediction time, given all the learned model parameters, we conduct an inference step to infer the paragraph vector as well as the label using MAP estimate over the test text.

The above SGPV may have limited modeling ability on text representation since it mainly relies on uni-grams. As we know, word order information is often critical in capturing the meaning of texts. For example, "machine learning" and "learning machine" are totally different in meaning with the same words. There has been a variety of deep models using complex architectures such as convolution layers or recurrent structures to help capture such order information at the expense of large computational cost.

In order to capture the word order information, we propose to extend SGPV by introducing an additional generative process for n-grams, so that we can incorporate the word order information into the model and meanwhile keep its simplicity in learning. We name this extension as SGPV-ngram. In practice, each n-gram is treated as a special token and is directly predicted by each text. Thus, texts including semantically similar word sequences will tend to be closer to each other in vector space. Here we take the generative process of SGPV-bigram as an example.

For each text $d_n \in D, n = 1, 2, \ldots, N$:
(a) Draw paragraph vector $d_n \sim \mathcal{N}(\mu, \Sigma)$
(b) For each word $w_i^n \in d_n, i = 1, 2, \ldots, l_n$:
 Draw word vector $w_i^n \sim \text{softmax}(d_n \cdot W)_i$
(c) For each bigram $g_i^n \in d_n, i = 1, 2, \ldots, s_n$:
 Draw bigram vector $g_i^n \sim \text{softmax}(d_n \cdot G)_i$
(d) Draw label $y_n | d_n, U, b \sim \text{softmax}(U \cdot d_n + b)$

where G denotes a $K \times S$ bigram embedding matrix with $G_{*j} = g_j$, and S denotes the size of bigram vocabulary. The joint probability over the whole corpus is then defined as

$$p(D) = \prod_{n=1}^{N} \int p(d_n | \mu, \Sigma) \Big(\prod_{w_i^n \in d_n} p(w_i^n | W, d_n) \Big) \Big(\prod_{g_i^n \in d_n} p(g_i^n | G, d_n) \Big) p(y_n | d_n, U, b) dd_n. \quad (6)$$

5 Experiments

In this section, we introduce the experimental settings and empirical results on a set of text classification tasks.

5.1 Dataset and Experimental Setup

We made use of five publicly available benchmark datasets in comparison.

- **TREC:** The TREC Question Classification dataset [12][1] which consists of 5,452 train questions and 500 test questions. The goal is to classify a question into 6 different types depending on the answer they seek for.

---

[1] http://cogcomp.cs.illinois.edu/Data/QA/QC/.

- **Subj:** Subjectivity dataset [16][2] which contains 5,000 subjective instances and 5,000 objective instances. The task is to classify a sentence as being subjective or objective.
- **MR:** Movie reviews [17] with one sentence per review. There are 5,331 positive sentences and 5,331 negative sentences. The objective is to classify each review into positive or negative category.
- **SST-1:** Stanford Sentiment Treebank [19][3]. SST-1 is provided with train/dev/test splits of size 8,544/1,101/2,210. It is a fine-grained classification over five classes: very negative, negative, neutral, positive, and very positive.
- **SST-2:** SST-2 is the same as SST-1 but with neutral reviews removed. We use the standard train/dev/test splits of size 6,920/872/1,821 for the binary classification task.

Preprocessing steps were applied to all datasets: words were lowercased, non-English characters and stop words occurrence in the training set are removed. For fair comparison with other published results, we use the default train/test split for TREC, SST-1 and SST-2 datasets. Since explicit split of train/test is not provided by Subj and MR datasets, we use 10-fold cross-validation instead.

In our model, text and word vectors are randomly initialized with values uniformly distributed in the range of $[-0.5, +0.5]$ with 300-dimensionality. Following the practice in [14], we set the noise distributions for context and words as $P_{\mathrm{nw}}(w) \propto \#(w)^{0.75}$. We adopt the same linear learning rate strategy where the initial learning rate of our models is 0.025. For unsupervised methods, we use support vector machines (SVM)[4] as the classifier.

5.2 Baselines

We adopted both unsupervised and supervised methods on text representation as baselines.

Unsupervised Baselines

- **Bag-of-word-TFIDF and Bag-of-bigram-TFIDF** [3]. In the Bag-of-Word-TFIDF scheme, each text is represented as the *tf-idf* value of chosen feature-words. The Bag-of-bigram-TFIDF model is constructed by selecting the most frequent unigrams and bigrams from the training subset. We use the vanilla TFIDF in the gensim library[5].
- **LSI** [2] and **LDA** [1]. LSI maps both texts and words to lower-dimensional representations using SVD decomposition. In LDA, each word within a text is modeled as a finite mixture over an underlying set of topics. We use the vanilla LSI and LDA in the gensim library with topic number set as 300.

---

[2] https://www.cs.cornell.edu/people/pabo/movie-review-data/.
[3] http://nlp.stanford.edu/sentiment/. We train the model on both phrases and sentences but only score on sentences at test time, as in [10].
[4] http://www.csie.ntu.edu.tw/~cjlin/libsvm/.
[5] http://radimrehurek.com/gensim/.

- **cBow** [14]. Continuous Bag-of-Words model. We use average pooling as the global pooling mechanism to compose a sentence vector from a set of word vectors. We use the result reported in [24].
- **PV** [10]. Paragraph Vector is an unsupervised model to learn distributed representations of words and paragraphs. We use the result reported in [24].
- **FastSent** [4]. In FastSent, given a simple representation of some sentence in context, the model attempts to predict adjacent sentences.

Note that unlike LDA and GPV, LSI, cBow, and FastSent cannot infer the representations of unseen texts. Therefore, these three models need to fold-in all the test data to learn representations together with training data, which makes it not efficient in practice.

Supervised Baselines

- **NBSVM** and **MNB** [23]. Naive Bayes SVM and Multinomial Naive Bayes with uni-grams and bi-grams.
- **DAN** [7]. Deep averaging network (DAN) uses the average of the word vectors as the input and applies multiple neural layers to learn text representation under supervision.
- **CNN** [9]. Convolutional Neural Networks (CNN) utilize layers with convolving filters for sentence modeling. We take the CNN-multichannel variant into consideration.
- **DCNN** [8]. Dynamic Convolutional Neural Network (DCNN) uses a convolutional architecture that replaces wide convolutional layers with dynamic pooling layers.
- **MV-RNN** [18]. Matrix-Vector RNN (MV-RNN) represents every word and longer phrase in a parse tree as both a vector and a matrix.
- **DRNN** [6]. Deep Recursive Neural Networks (DRNN) is constructed by stacking multiple recursive layers.
  Dependency Tree-LSTM [21]. The Dependency Tree-LSTM based on LSTM structure uses dependency parses of each sentence.

5.3   Performance of Generative Paragraph Vector

We first evaluate the GPV model by comparing with the unsupervised baselines on the TREC, Subj and MR datasets. As shown in Table 1, GPV can also outperform all the baselines on three tasks. This indicates that GPV can capture better semantic representations of texts by using a complete generation process. Specially, GPV works better than PV over the three tasks, which demonstrates the benefits of introducing a corpus-wide constraint (i.e., the prior distribution) over the paragraph vectors. Compared with topic models (*i.e.*, LDA), unlimited hidden topics (*i.e.*, word clusters) in GPV helps to model the texts while the explicit "topic" layer may lead to non-optimal word clustering. Moreover, the results show that for unsupervised text representation, bag-of-words representation is quite simple yet powerful which can beat many embedding models.

(a) Subj   (b) MR   (c) TREC

Fig. 4. PCA is used to reduce the dimension to two for visualization purposes. Different colors correspond to different objective classes. (Color figure online)

Table 1. Classification accuracies of GPV compared with other unsupervised models.

Model	TREC	Subj	MR
BoW-TFIDF	87.2	89.8	76.7
Bigram-TFIDF	89.6	90.9	76.1
LSI	79.2	89.4	69.4
LDA	70.6	79.8	65.2
cBow	87.3	91.3	77.2
PV	91.8	90.5	74.8
FastSent	76.8	88.7	70.8
GPV	**92.2**	**92.1**	**77.9**

We further visualized the learned text representations from GPV to get some intuitive understanding of the model. We used PCA to project the learned paragraph vectors into a 2-dimension space, and the results are shown in Fig. 4. From the results we can see that GPV are able to capture the semantic relatedness between texts, even without the usage of the class labels. The results indicate that GPV might be used as an good feature extractor for downstream applications.

5.4   Performance of Supervised Generative Paragraph Vector

We compare SGPV model to supervised baselines on all the five classification tasks. Empirical results are shown in Table 2. We can see that SGPV achieves comparable performance against other deep learning models. Note that SGPV is much simpler than these deep models with significantly less parameters and no complex structures. Moreover, deep models with convolutional layers or recurrent structures can potentially capture compositional semantics (*e.g.*, phrases), while SGPV only relies on uni-gram. In this sense, SGPV is quite effective in learning text representation. Meanwhile, if we take Table 1 into consideration, it is not surprising to see that SGPV can consistently outperform GPV on all the three classification tasks. This also demonstrates that it is more effective to directly fit

Table 2. Classification accuracies of our SGPV models against other supervised representation models. For all the supervised baselines, we use the results reported in the corresponding papers.

Model	SST-1	SST-2	TREC	Subj	MR
NBSVM	-	-	-	93.2	79.4
MNB	-	-	-	**93.6**	79.0
DAN	47.7	86.3	-	-	-
CNN-multichannel	47.4	88.1	92.2	93.2	**81.1**
DCNN	48.5	86.8	93.0	-	-
MV-RNN	44.4	82.9	-	-	79.0
DRNN	49.8	86.6	-	-	-
Dependency Tree-LSTM	48.4	85.7	-	-	-
SGPV	45.2	86.3	92.6	92.4	79.2
SGPV-bigram	**55.9**	**91.8**	**93.2**	**93.6**	79.8

supervised representation models than to learn a general purpose representation in prediction scenarios.

By introducing bi-grams, SGPV-bigram can outperform all the other deep models on four tasks. In particular, the improvements of SGPV-bigram over other baselines are significant on SST-1 and SST-2. These results again demonstrated the effectiveness of our proposed SGPV model on text representations. It also shows the importance of word order information in modeling text semantics.

6   Conclusions

In this paper, we introduce GPV and SGPV for learning distributed representations for pieces of texts. With a complete generative process, our models are able to leverage the corpus-wide information as well as labels to help text representation learning. Our models keep as simple as PV models, and thus can be efficiently learned over large scale text corpus. Even with such simple structures, both GPV and SGPV can produce state-of-the-art results as compared with existing baselines, especially those complex deep models. For future work, we may consider other probabilistic distributions for both paragraph vectors and word vectors.

Acknowledgments. This work was funded by the 973 Program of China under Grant No. 2014CB340401, the National Natural Science Foundation of China (NSFC) under Grants No. 61425016, 61472401, 61722211, and 20180290, the Youth Innovation Promotion Association CAS under Grants No. 20144310, and 2016102, and the National Key R&D Program of China under Grants No. 2016QY02D0405.

References

1. Blei, D.M., Ng, A.Y., Jordan, M.I.: Latent Dirichlet allocation. J. Mach. Learn. Res. **3**(Jan), 993–1022 (2003)
2. Deerwester, S., Dumais, S.T., Furnas, G.W., Landauer, T.K., Harshman, R.: Indexing by latent semantic analysis. J. Am. Soc. Inf. Sci. **41**(6), 391 (1990)
3. Harris, Z.S.: Distributional structure. Word **10**(2–3), 146–162 (1954)
4. Hill, F., Cho, K., Korhonen, A.: Learning distributed representations of sentences from unlabelled data. arXiv preprint arXiv:1602.03483 (2016)
5. Hofmann, T.: Probabilistic latent semantic indexing. In: SIGIR, pp. 50–57. ACM (1999)
6. Irsoy, O., Cardie, C.: Deep recursive neural networks for compositionality in language. In: Advances in Neural Information Processing Systems, pp. 2096–2104 (2014)
7. Iyyer, M., Manjunatha, V., Boyd-Graber, J., Daumé III, H.: Deep unordered composition rivals syntactic methods for text classification. In: ACL (2015)
8. Kalchbrenner, N., Grefenstette, E., Blunsom, P.: A convolutional neural network for modelling sentences. arXiv preprint arXiv:1404.2188 (2014)
9. Kim, Y.: Convolutional neural networks for sentence classification. arXiv preprint arXiv:1408.5882 (2014)
10. Le, Q.V., Mikolov, T.: Distributed representations of sentences and documents. ICML **14**, 1188–1196 (2014)
11. Levy, O., Goldberg, Y., Dagan, I.: Improving distributional similarity with lessons learned from word embeddings. Trans. Assoc. Comput. Linguist. **3**, 211–225 (2015)
12. Li, X., Roth, D.: Learning question classifiers. In Proceedings of the 19th International Conference on Computational Linguistics, vol. 1, pp. 1–7. Association for Computational Linguistics (2002)
13. Mcauliffe, J.D., Blei, D.M.: Supervised topic models. In: NIPS, pp. 121–128 (2008)
14. Mikolov, T., Chen, K., Corrado, G., Dean, J.: Efficient estimation of word representations in vector space. arXiv preprint arXiv:1301.3781 (2013)
15. Mikolov, T., Sutskever, I., Chen, K., Corrado, G. S., Dean, J.: Distributed representations of words and phrases and their compositionality. In: Advances in Neural Information Processing Systems, pp. 3111–3119 (2013)
16. Pang, B., Lee, L.: A sentimental education: sentiment analysis using subjectivity summarization based on minimum cuts. In: ACL, p. 271. Association for Computational Linguistics (2004)
17. Pang, B., Lee, L.: Seeing stars: exploiting class relationships for sentiment categorization with respect to rating scales. In: ACL, pp. 115–124. Association for Computational Linguistics (2005)
18. Socher, R., Huval, B., Manning, C.D., Ng, A.Y.: Semantic compositionality through recursive matrix-vector spaces. In: EMNLP, pp. 1201–1211. Association for Computational Linguistics (2012)
19. Socher, R., et al.: Recursive deep models for semantic compositionality over a sentiment TreeBank. In: EMNLP, vol. 1631, p. 1642. Citeseer (2013)
20. Sutskever, I., Vinyals, O., Le, Q. V.: Sequence to sequence learning with neural networks. In: NIPS, pp. 3104–3112 (2014)
21. Tai, K.S., Socher, R., Manning, C.D.: Improved semantic representations from tree-structured long short-term memory networks. arXiv preprint arXiv:1503.00075 (2015)

22. Tellex, S., Katz, B., Lin, J., Fernandes, A., Marton, G.: Quantitative evaluation of passage retrieval algorithms for question answering. In: SIGIR, pp. 41–47. ACM (2003)
23. Wang, S., Manning, C.D.: Baselines and bigrams: simple, good sentiment and topic classification. In: ACL, pp. 90–94. Association for Computational Linguistics (2012)
24. Zhao, H., Lu, Z., Poupart, P.: Self-adaptive hierarchical sentence model. arXiv preprint arXiv:1504.05070 (2015)

Generative Adversarial Graph Representation Learning in Hyperbolic Space

Xiaomei Liu[1,2], Suqin Tang[1,2(✉)], and Jinyan Wang[1,2]

[1] Guangxi Key Lab of Multi-Source Information Mining and Security,
Guangxi Normal University, Guilin 541004, China
17602668@qq.com
[2] College of Computer Science and Information Technology,
Guangxi Normal University, Guilin 541004, China

Abstract. Representation learning can provide a compact representation of features. There are a large number of representation learning methods which have been successfully applied to feature learning of graph structured data. However, most of the existing graph representation learning methods does not consider the latent hierarchical structure of the data. They mainly focus on high-dimensional Euclidean space for learning. Recent studies have shown that graph structured data are suitable for being embedded in hyperbolic space and that hyperbolic space can be naturally equipped to model hierarchical structures where they outperform Euclidean embeddings. Therefore, we comprehensively consider the hierarchical structure characteristics of graphs, and learn the vector representation of nodes in hyperbolic space by using the principle of generative adversarial learning. By using two models to simultaneously capture hierarchy and similarity and let them compete with each other, the performance of learning is alternately boosted. In this paper, node classification, link prediction and visualization are performed on multiple public datasets. The results show that this method performs well in multiple tasks.

Keywords: Node representation · Hyperbolic geometry
Generative adversarial · Feature learning

1 Introduction

Graphs (also called networks) are often found in important and emerging domains. For example, social networks, molecular graph, biological protein networks, knowledge graphs and other fields can be easily modeled as graphs. The information that hidden in the graphs can be found by analyzing these graphs. Therefore, effective graph analysis can benefit to many applications for decades, such as node classification, link prediction, community detection and recommendation systems [1]. However, in order to extract structured information from the graph, traditional machine methods often rely on correlation statistics (e.g., degree or clustering coefficient), kernel functions, or well-designed features of the graph to measure the local neighborhood structure. Designing these features is a time consuming and expensive process. Recently, there are a lot of methods that try to learn to encode the representation of the structured information

© Springer Nature Switzerland AG 2018
S. Zhang et al. (Eds.): CCIR 2018, LNCS 11168, pp. 119–131, 2018.
https://doi.org/10.1007/978-3-030-01012-6_10

about the graph. The idea behind the representation learning methods is to learn the mapping of nodes or entire (sub)graphs into low-dimensional vector space [2]. In recent years, deep learning technology has achieved great success in many areas, such as speech processing, image recognition and natural language processing. Many graph representation learning methods have been successively proposed, such as DeepWalk [3], LINE [4], node2vec [5] and so on.

In the existing graph representation learning, few methods comprehensively consider the complex network structure of the graph. Almost previous work explicitly or implicitly assumed that the vector space is Euclidean space (using Euclidean dot product) on neural embedding [6]. Recent work in the field of complex networks has found that the hyperbolic geometric model can well reveal complex network features, like power law distributions. Many real-world objects can be organized according to a latent hierarchy, and they exhibit a kind of latent tree-like structure. This hyperbolic space can be seen as a continuous version of the tree. With basing on these related studies, Nickel et al. [7] embed symbolic data into a n-dimensional Poincaré ball for learning hierarchical representations and show experimentally that Poincaré embeddings can outperform Euclidean embeddings significantly on data with latent hierarchies. Tay et al. [8] propose a simple neural network that learns QA embeddings in hyperbolic space. Chamberlain et al. [6] propose learning neural embeddings of graphs in hyperbolic space which based on Skip-Gram [9].

In order to improve the existing representation learning methods, we propose a modified graph representation learning method from the perspective of combining complex networks and graph data mining. We call it Adversarial Network Embedding in Hyperbolic Space - HyperANE. This method draws on the hyperbolic distance metric proposed by Chamberlain et al. [6] and the graph representation learning framework proposed by Wang et al. [10]. In this paper, the main work is as follows:

- Assuming that the graph data has a latent hierarchical structure, and the semantic similarity of nodes reflects their distance in the embedded space. By transforming the distance metric from Euclidean space to hyperbolic space, this paper explicitly captures the hierarchical structure features of data in the embedded space.
- Unsupervised representation learning method is used for learning a robust node representation. Using a generative model and an adversarial model to simultaneously preserve the network's hierarchy and node similarity, and alternately boost their performance through their mutual competition.
- The effectiveness of the graph representation learning method is verified by performing node classification, link prediction and visualization experiments on multiple public datasets. Compared with some existing graph representation learning methods, the results have a good performance in multiple tasks.

The rest of this paper is organized as follows. In Sect. 2, we briefly review related work, mainly involving graph representing learning methods and hyperbolic geometry in complex networks. In the Sect. 3, the generative adversarial graph representation learning in the hyperbolic space is introduced in detail. In the Sect. 4, the proposed algorithm is evaluated on several different tasks, and the relevant experimental results are analyzed.

2 Related Work

In general, graph representation learning aims to preserve the graph attribute information in a low-dimensional space as much as possible, that is, by embedding the graph in a low-dimensional vector space and encoding the feature information of the graph with a low-dimensional vector. The difference between the different graph representation learning algorithms mainly lies in how they define the graph properties to preserve [1]. Early graph decomposition based on matrix decomposition represent graph attributes in a matrix form (e.g., node pairwise similarity) and factorize the matrix to obtain node embedding. The problem of graph embedding can be seen as a dimensionality reduction method for holding structures in hypothetical input data in low-dimensional manifolds. For example, MDS [11], Isomap [12], LLE [13] and LE [14] are based on matrix decomposition. However, the construction of similarity matrix or feature decomposition is time-consuming and space-consuming, making matrix decomposition inefficient and unable to adapt to large-scale graphs. Recently, the word representation learning method has received extensive attention in the field of natural language processing. Inspired by this, DeepWalk [3] uses a neural-language model (Skip-Gram [9]) for graph embedding. DeepWalk's success has inspired many follow-up studies that apply deep learning models to graph representation learning. Most studies follow the idea of DeepWalk, but they change the sampling method of random walk or the similarity measure between nodes. By defining the graph in the spectral domain, convolutional neural networks (CNNs) and their variants are also widely used for graph embedding [1]. More studies of graph representation learning can be found in [1, 2, 15].

Hyperbolic geometry does not follow the fifth hypothesis of Euclidean geometry (parallel hypothesis). Unlike Euclidean geometry, hyperbolic geometry belongs to non-Euclidean geometry which studies spaces of constant negative curvature. In the hyperbolic space, even though the infinite tree has approximately isometric embedding, it has been successfully used to model a complex network with hierarchical structure, power law degree distribution and high clustering [16]. An obvious feature of hyperbolic space is that it is in a sense larger than the more familiar Euclidean space; the circumference or area of a circle grows exponentially with its radius, not a polynomial. Informally, trees can therefore be thought of as "discrete hyperbolic spaces". Formally, trees, even infinite ones, allow nearly isometric embeddings into hyperbolic spaces [16]. This shows that low-dimensional hyperbolic space can provide effective representations of data in a way that low-dimensional Euclidean space cannot. However, this makes the hyperbolic space difficult to visualize because even 2D hyperbolic plane cannot be isometric embedded into any dimension of Euclidean space. For this reason, hyperbolic space is represented in many different ways. Each representation preserves some geometrical features and distorts others [6]. Although both volume and distance are distorted, the Poincaré disk model is conformal, which means that the Euclidean and hyperbolic angles between the lines are equal. Herein, we use the Poincaré disk model of hyperbolic space to describe.

3   Adversarial Embedding in Hyperbolic Space

In the following, we are interested in finding embeddings of symbolic data such that their distance in the embedding space reflects their semantic similarity [7]. We assume that the graph data has a latent hierarchical structure. Following Chamberlain et al. [6], we will use the conformal properties and circular symmetry of the Poincaré disk. And, we propose a Generative Adversarial Networks (GANs) [17] architecture for learning node embedding vectors in hyperbolic space which is called HyperANE. An overview of our model's architecture can be seen in Fig. 1.

A graph can be denoted as $\mathcal{G} = \mathcal{V}, \mathcal{E}$, where $\mathcal{V} = \{v_1, \ldots, v_V\}$ represents a set of nodes and $\mathcal{E} = \{e_{ij}\}_{i,j=1}^{V}$ represents a set of edges. For a given node v_c, define $\mathcal{N}(v_c)$ as the set of nodes directly connected to v_c. The conditional probability $\mathcal{P}_{true}(v|v_c)$ represents the real connection distribution of node v_c that reflects the distribution preference of all other nodes in $\mathcal{V}$ and v_c. $\mathcal{N}(v_c)$ can be viewed as a set of observations from $\mathcal{P}_{true}(v|v_c)$. It is well known that training traditional GANs is subtle and unstable, while Wasserstein GAN (WGAN) can prevent mode collapse and lead to more stable training overall. So unlike GraphGAN [10], we train our model based on the Wasserstein GAN with gradient penalty (WGAN-GP) [18]. WGAN-GP with a gradient penalty term can mitigate the instability caused by the weight clipping in WGAN. Like any typical GANs architecture, HyperANE consists of two main components - a generator G and a discriminator D. The critic loss is as follows:

$$
\max_\theta \min_w V(G, D) = \sum_{c=1}^{V} \left(\mathbb{E}_{\tilde{v} \sim G(\cdot|v_c;\theta)}[D_w(\tilde{v}, v_c)] - \mathbb{E}_{v \sim \mathcal{P}_{true}(\cdot|v_c)}[D_w(v, v_c)] + \lambda \mathbb{E}_{\hat{v} \sim \mathcal{P}(\hat{v})} \left[\left(\|\nabla_{\hat{v}} D_w(\hat{v}, v_c)\|_2 - 1 \right)^2 \right] \right)
\tag{1}
$$

Given a sample (v, v_c), the discriminator will try to approximately fit the Wasserstein distance and wants to pull as high as possible the scores of the real sample v from the true connection distribution $\mathcal{P}_{true}(v|v_c)$ and pull down the score of the fake sample $\tilde{v}$ from the generator $G(\tilde{v}|v_c; \theta)$. The generator approximately minimizes the Wasserstein distance between the true distribution and the generated distribution, that is, $G(\tilde{v}|v_c; \theta)$ tries to perfectly fit $\mathcal{P}_{true}(v|v_c)$ and generate the node $\tilde{v}$ that is most likely to be connected to v_c. Finally, the gradient punishes the joint $\hat{v}$ of the true and fake samples. In the rest, we describe each stage of this process to make our design be more detailed.

3.1   Discriminator

In the discriminator D_w, we try to approximately fit the Wasserstein distance using a ReLU-MLP three-layer neural network with 512 hidden units. Then, we only need to update w by gradient descent.

3.2   Generator

In contrast to the discriminator, the generator aims to pull the score of the fake sample as high as possible. In other words, given the latent variable z, the generator changes its approximate connection distribution (via its parameter θ) to increase the score of the sample which it generated. Again, we need to update θ with gradient descent:

$$\nabla_\theta V(G, D) = -\mathbb{E}_{z \sim \mathcal{P}(z)}[\nabla_\theta D_w(G_\theta(z))] \tag{2}$$

In HyperANE, in order to make better use of the hyperbolic geometric character-istics of complex networks in the generator and discriminator, the node is uniformly embedded in the hyperbolic space. Therefore, the distance or similarity of the two input nodes from the Euclidean inner product is changed to hyperbolic inner product. In order to greatly simplify its mathematical form, we use the symmetry of the model and the nodes are described in Poincaré disk using polar coordinates, e.g. $x = (r_h, \theta_h)$, where $r_h \in [0,1)$ and $\theta_h \in [0,2\pi)$. In the hyperbolic coordinate system, the distance of $x = (r_h, \theta_h)$ to the origin is $r_h = 2 \operatorname{arctanh} r_e$. In a Poincaré disk, the inner product of two vectors $x = (r_x, \theta_x)$ and $y = (r_y, \theta_y)$ is given by:

$$\langle x, y \rangle = \|x\|\|y\| \cos(\theta_x - \theta_y) \tag{3}$$

$$= 4 \operatorname{arctanh} r_x \operatorname{arctanh} r_y \cos(\theta_x - \theta_y) \tag{4}$$

Next, we discuss the implementation of G. As mentioned earlier, the entire network is embedded in the hyperbolic space, and we use the hyperbolic inner product to calculate the distance or similarity. A simple way is to define the generator as a softmax function for all other nodes:

$$G(\tilde{v}|v_c; \theta) = \frac{exp(\langle \mathbf{g}_{\tilde{v}}, \mathbf{g}_{v_c} \rangle)}{\sum_{\tilde{v} \neq v_c} exp(\langle \mathbf{g}_{\tilde{v}}, \mathbf{g}_{v_c} \rangle)} \tag{5}$$

Where $\mathbf{g}_{\tilde{v}}, \mathbf{g}_{v_c} \in \mathbb{R}^k$ is the k-dimensional vector representation of nodes $\tilde{v}$ and v_c of generator G, respectively, and θ is union of all $\mathbf{g}_{v_c}$.

In this case, in order to update θ in each iteration, the approximate connected distribution $G(\tilde{v}|v_c; \theta)$ is calculated according to Eq. (5), and generator G randomly samples a group of samples $(\tilde{v}, v_c)$, and update θ with a random gradient descent. However, the calculation of softmax involves all nodes in the graph, and although some of the popular softmax alternatives, such as hierarchical softmax [19] and negative sampling [19], can reduce the amount of computation in a certain extent, they do not consider the structure information of the graph. So, both of them cannot achieve satisfactory performance when applied to graph representation learning [10].

In order to solve the above problem, we borrow a softmax alternative called graphsoftmax [10] which proposed in GraphGAN. The method firstly constructs the breath-first search (BFS) tree of the original graph, and it adopts an online sampling strategy of random walk. The calculation of softmax involves only the neighbor nodes of the node, which greatly reduces the amount of computation and takes into account the structure information of the graph. Unlike the sampling method in GraphGAN, considering the homophily and structural equivalence of the graph, we firstly use a flexible neighborhood sampling strategy - node2vec [5]. The method uses a return parameter p and an input-output parameter q to control a second-order Markov random walk process. By the node2vec sampling in the original graph, the HyperANE first obtains the generated sample of the node sequence, and then takes the starting node v_c of each sequence as the root node. The starting node v_c and the remaining nodes in the sequence constitute the sampling node pairwise $(v_c, \tilde{v})$. HyperANE begins the breadth-first search on the original graph $\mathcal{G}$ with starting v_c, which provides us a BFS tree T_c with rooted at v_c. For a given node v_c and its neighbor $v_i \in \mathcal{N}_c(v)$, the relevance probability of defining a given node v is:

$$P_c(v_i|v) = \frac{exp(\langle \mathbf{g}_{v_i}, \mathbf{g}_v \rangle)}{\sum_{v_j \in \mathcal{N}_c(v)} exp(\langle \mathbf{g}_{v_j}, \mathbf{g}_v \rangle)} \tag{6}$$

Actually, this is a softmax function on $\mathcal{N}_c(v)$. To compute $G(\tilde{v}|v_c; \theta)$, here each node $\tilde{v}$ can be reached via a unique path in the BFS-tree T_c with rooted at v_c. The path searched in the BFS-tree is represented as $P_{v_c \to \tilde{v}} = (v_{r0}, v_{r1}, \ldots, v_{rm})$, where $v_{r0} = v_c$ and $v_{rm} = \tilde{v}$. Then, the calculated $G(\tilde{v}|v_c; \theta)$ is defined as follows:

$$G(\tilde{v}|v_c) \triangleq \prod_{j=1}^{m} P_c(v_{r_j}|v_{r_{j-1}}) \tag{7}$$

Where $P_c(\cdot|\cdot)$ is the relevance probability defined in Eq. (6).

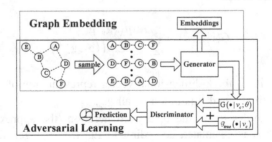

Fig. 1. The framework of HyperANE.

3.3 Gradient Update

HyperANE embeds a graph in hyperbolic space, using Poincaré disk instead of Euclidean vector space. The embedding is learned by optimizing an objective function that learns the vector representation of the nodes by WGAN-GP adversarial training, where HyperANE replaces the Euclidean inner product with the hyperbolic inner product in the softmax functions of the generator.

Generative adversarial networks use the back-propagation to learn model. The simplest way to perform back-propagation is to use the natural hyperbolic coordinates on the disk and map back to Euclidean coordinates only at the end. Assume that v_c and $\tilde{v}$ are the vector representations of the nodes v_c and $\tilde{v}$,so that the hyperbolic inner product is $u = \langle v_c, \tilde{v} \rangle = r_{v_c} r_{\tilde{v}} \cos(\theta_{\tilde{v}} - \theta_{v_c})$. The main drawback of this coordinate system is that it introduces a singularity at the origin. In order to solve the complexity caused by the radius being less than or equal to zero, we initialize all vectors to be in a patch of space that is small relative to its distance from the origin [6]. Taking the derivatives yields:

$$\frac{\partial V(G,D)}{\partial r_{v_c}} = \frac{\partial V(G,D)}{\partial u} \frac{\partial u}{\partial r_{v_c}} = \frac{\partial V(G,D)}{\partial u} r_{\tilde{v}} \cos(\theta_{\tilde{v}} - \theta_{v_c}) \tag{8}$$

$$\frac{\partial V(G,D)}{\partial \theta_{v_c}} = \frac{\partial V(G,D)}{\partial u} \frac{\partial u}{\partial \theta_{v_c}} = \frac{\partial V(G,D)}{\partial u} r_{v_c} r_{\tilde{v}} \sin(\theta_{\tilde{v}} - \theta_{v_c}) \tag{9}$$

The Jacobian is then

$$\partial V = \frac{\partial V}{\partial r_h} \widehat{r_h} + \frac{1}{\sinh r_h} \frac{\partial V}{\partial \theta_h} \widehat{\theta_h} \tag{10}$$

Calculating the derivative of another node vector follows the same pattern. After the backward propagation is complete, the vectors in the Poincaré disk map back to Euclidean coordinates through $\theta_h \rightarrow \theta_e$ and $r_h \rightarrow \tanh \frac{r_h}{2}$.

Table 1. Statistical information for the datasets.

Datasets	Cora	Citeseer	Wiki	Blogcatalog	Flickr
#Nodes	2078	3312	2405	10312	80513
#Edges	5429	4732	17981	333983	5899882
# categories	7	6	19	39	195

4   Experiment

Graph representation learning is beneficial for various graph analysis applications because the vector representation can be efficiently processed in time and space. In this section, HyperANE will be applied to some real-world scenarios.

4.1   Experiment Settings

Datasets. Here, we conduct experiments on five real-world datasets and regard these networks as undirected graphs. After removing the self-loops and isolated nodes, the relevant statistics of each dataset are shown in Table 1. Cora [20] and Citeseer [20] are paper citation networks. Wiki [21] is a network with nodes as web pages and edges as the hyperlinks between them. It is denser than Cora and Citeseer. Blogcatalog [22] is a network of social relationships of the bloggers listed on the BlogCatalog website. The labels represent blogger interests provided by the bloggers. Flickr [22] is a network of the contacts between users of the photo sharing website. The labels represent the interest groups of the users. Different from Cora, Citeseer and Wiki, each user may be labeled with more than one category in Blogcatalog and Flickr.

Baselines. The baselines chosen in this paper mainly use the topology features of the nodes, and it does not involve the external information of the nodes. We chose the methods commonly used in other literatures, namely:

- SpectralCluster [23] uses the normalized Laplacian $\mathcal{L}$ of graph $\mathcal{G}$. The representation in $\mathbb{R}^d$ is generated from the d smallest eigenvectors of $\mathcal{L}$. Utilizing $\mathcal{L}'$s eigenvectors implicitly assumes that graph partitioning is useful for classification.
- DeepWalk [3] first transforms the network into a sequence of nodes by truncating random walks, then enters the Skip-Gram model to learn the node representation vectors.
- LINE [4] respectively maintains the first-order and second-order proximity of the graph by modeling the co-occurrence and conditional probabilities between the nodes.
- Node2vec [5] is a variation of DeepWalk that designs a biased random walk process to explore the neighbors of the node.
- GraphGAN [10] designs a generative adversarial network to train node embedding and proposes graph softmax as the implementation of the generator.

Parameter Settings. In HyperANE, the gradient penalty coefficient λ is set to 10 and the Adam optimization algorithm's parameter α is set to 0.0001, β_1 is set to 0, and β_2 is set to 0.9. In each iteration, the number of samples corresponding to each node is set to 20. The number of critic iterations per generator iteration is set to 5. The node2vec sampling in the HyperANE is set to a random walk step length of 5 for each node, and each node is sampled 4 times. For the rest of hyper-parameters, we use default parameter values as suggested by their original papers. We fix the representation dimension of the node on 128 dimensions (same for all baseline methods).

Table 2. Accuracy (%) of multi-class classification on Cora, Citeseer and Wiki.

Datasets	Cora			Citeseer			Wiki		
%Labeled Nodes	20%	50%	80%	20%	50%	80%	20%	50%	80%
SpectralCluster	58.78	70.05	72.16	46.83	53.77	54.67	53.03	55.95	58.96
DeepWalk	68.83	72.79	74.17	52.68	56.44	58.04	57.22	62.96	64.63
LINE	69.50	72.68	74.00	51.04	55.10	57.08	57.63	60.25	64.86
Node2vec	71.49	74.52	76.28	53.47	56.81	58.34	61.52	65.04	68.14
GraphGAN	72.45	76.64	**78.52**	55.52	58.34	60.63	**61.66**	66.69	69.68
HyperANE	**72.50**	**77.28**	78.50	**56.71**	**59.25**	**60.87**	61.14	**67.02**	**71.60**

4.2 Node Classification

Node classification is to assign one or more class labels to each node in the graph. We treat node embeddings as node features and feed them into a classifier. Example classifiers include SVM, logistic regression and k-nearest neighbor classification. Then given the embedding of unlabeled nodes, the trained classifier can predict the labels.

We randomly sample different proportions of the labeled nodes from the dataset. They are used as training data and the rest as test data. This process is repeated 10 times and we report average performance. Specifically, we calculated the proportion that the classifier can correctly predict for Cora, Citeseer and Wiki, that is accuracy. Since Blogcatalog and Flickr have been tagged with multiple labels, we calculate Macro-F1 and Micro-F1 for them. For all models, we use the one-vs-rest SVM in LibLinear [24] package to predict the most likely labels for each node. The experimental results are shown in Tables 2 and 3. The experimental results show that as the training data increases, the performance of the classifier is improved. In most of the results, HyperANE has a good performance. Compared with small-scale datasets, the chosen methods in this paper performed poorly in Blogcatalog and Flickr. Studies have shown that most of the representational learning algorithms still lack performance on large scale datasets, and it is still an urgent issue to be solved in terms of how to speed up learning rate and improve performance.

Table 3. Macro-F1 and Micro-F1 (%) of multi-label classification on Blogcatalog and Flickr.

Metrics	Datasets	Blogcatalog			Flickr		
	%Labeled Nodes	20%	50%	80%	2%	5%	10%
Micro-F1(%)	SpectralCluster	23.2	32.9	34.5	5.1	20.2	28.5
	DeepWalk	29.8	35.9	38.3	7.9	32.9	35.5
	LINE	21.5	27.6	30.2	5.5	30.3	32.6
	Node2vec	31.0	37.4	39.9	9.8	32.0	36.0
	GraphGAN	32.8	38.6	40.9	10.3	30.0	35.8
	HyperANE	**35.8**	**40.2**	**41.5**	**11.2**	**33.6**	**37.1**
Maro-F1(%)	SpectralCluster	13.2	15.3	15.7	1.5	5.7	9.2
	DeepWalk	18.7	20.9	21.5	3.1	13.8	15.7
	LINE	8.3	10.2	10.9	2.5	10.2	11.2
	Node2vec	20.4	23.2	23.6	3.5	14.7	17.0
	GraphGAN	21.5	22.9	24.6	5.2	17.5	**18.5**
	HyperANE	**22.3**	**24.8**	**26.0**	**6.1**	**18.1**	18.2

4.3 Link Prediction

Link prediction is intended to predict the existence of links. Intuitively, the more similar the two nodes are, the more likely there is an edge between them. For the purpose of link prediction, we need to score each pair of nodes given their embeddings, for example, cosine similarity, inner product, inverse $L2$-distance and so on.

First of all, we remove a portion of the existing links from the network. When only a few edges (such as 5%) are reserved as training data, the results of all methods are poor, because most of the nodes are isolated. So, we randomly remove 20% edges from Cora, Citeseer, Wiki, and 50% edges from Blogcatalog and Flickr. The vector representation of the node is learned based on the remaining networks. The pairs of nodes in the deleted edges are considered as positive samples, the same number of unconnected node pairs are randomly sampled as negative samples, and the positive and negative samples form a balanced dataset. Given the pair of nodes in the sample, we calculate the score based on the cosine similarity of the representation vector. The AUC is used to indicate the probability that the score of the unobserved link is higher than that of the link that does not exist. We also select Common Neighbors [25] as a baseline method which simply counts the common neighbors of two nodes as similarity score, because it has proven to be a simple and effective method. The experimental results are shown in Fig. 2. In the task of link prediction, HyperANE has slightly better results than other methods. For the dense Wiki and the large-scale Blogcatalog and Flickr datasets, the proposed method shows good and stable results, which indicates that the proposed method has strong flexibility and robustness. The traditional link prediction method - Common Neighbors, which doesn't belong to representation learning, also shows its simplicity and effectiveness.

4.4 Visualization

Network visualization is an indispensable method for analyzing high-dimensional data. It can intuitively help reveal the internal structure of data. Network visualization draws nodes in two-dimensional space with different colors. Different colors indicate different types of nodes. It vividly demonstrates that nodes belonging to the same category are close to each other.

Firstly, different embedding methods are used to map the Cora citation network to a low-dimensional space, and then the t-SNE [26] is used to further map the node's low-dimensional vectors to a two-dimensional space. Figure 3 compares the visualization results of different embedding methods. In the visualization task, the nodes of black category are grouped together by SpectralCluster, while other different categories of papers are mixed in the graph. For DeepWalk, LINE and Node2vec, although different categories of papers are clustered together, the boundaries between different clusters are not clear. Visualization of GraphGAN and HyperANE has formed seven major clusters. Although the boundaries between clusters are not very clear, the clusters are not as chaotic as Node2vec. Among them, in the cluster center, HyperANE's cluster boundaries are clearer than GraphGAN.

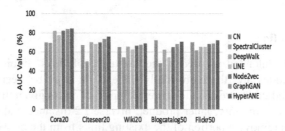

Fig. 2. Experimental results of link prediction.

(a) SpectralCluster (b) DeepWalk   (c) LINE   (d) Node2vec   (e) GraphGAN   (f) HyperANE

Fig. 3. Visualization results of Cora.

4.5 Effects of HyperANE Embedding Size

In order to evaluate how the change of the dimension represented by the HyperANE affects the performance of the classification task, we conduct experiments on the classification tasks of Cora, Citeseer and Wiki. For the sake of brevity, the parameters used in the HyperANE were used directly in this paper, 50% was selected as the training sample, and then only the number of dimensions is changed. Figure 4 shows the effect of different dimensions on the model. The performance between Cora, Citeseer and Wiki is very consistent. When the dimensions are better around 128-dimension, the accuracy is decreased when the dimension is too low or too high.

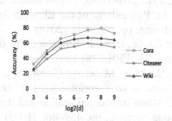

Fig. 4. Dimension sensitivity study.

5   Conclusion

In this paper, we propose a graph representation learning approach to preserve the hierarchical structure and similarity of the nodes in the graph. We explicitly embed the graph into the hyperbolic space, and we use the generator and discriminator to conduct the adversarial training to learn the robust node representations. By applying the learned node representations to node classification, link prediction and visualization, with some existing graph representation learning methods are compared, the results have good performance in multiple tasks. In the future, exploring the application scenarios that benefit from graph representation learning is a very important direction because it provides effective solutions for traditional problems from different perspectives.

Acknowledgements. This paper was supported by the National Natural Science Foundation of China (No.61662007, 61502111, 61763003), Guangxi science and technology (1598010-6), Guangxi Natural Science Foundation (Nos. 2016GXNSFAA380192) and Guangxi Collaborative Innovation Center of Multisource Information Integration and Intelligent Processing.

References

1. Cai, H., Zheng, V.W., Chang, K.: A comprehensive survey of graph embedding: Problems, techniques and applications. IEEE Trans. Knowl. Data Eng. **PP**(99), 1 (2018). https://doi.org/10.1109/tkde.2018.2807452
2. Hamilton, W.L., Ying, R., Leskovec, J.: Representation learning on graphs: methods and applications. arXiv preprint arXiv:1709.05584 (2017)
3. Perozzi, B., Al-Rfou, R., Skiena, S.: DeepWalk: online learning of social representations. In: Proceedings of the 20th ACM SIGKDD International Conference on Knowledge Discovery and Data Mining, pp. 701–710 (2014)
4. Tang, J., Qu, M., Wang, M., Zhang, M., et al.: LINE: large-scale information network embedding. In: Proceedings of the 24th International Conference on World Wide Web, pp. 1067–1077 (2015)
5. Grover, A., Leskovec, J.: Node2vec: scalable feature learning for networks. In: Proceedings of the 22nd ACM SIGKDD International Conference on Knowledge Discovery and Data Mining, pp. 855–864 (2016)
6. Chamberlain, B.P., Clough, J., Deisenroth, M.P.: Neural embeddings of graphs in hyperbolic space. arXiv preprint arXiv:1705.10359 (2017)
7. Nickel, M., Kiela, D.: Poincaré embeddings for learning hierarchical representations. In: Advances in Neural Information Processing Systems, pp. 6341–6350 (2017)
8. Tay, Y., Tuan, L.A., Hui, S.C.: Hyperbolic representation learning for fast and efficient neural question answering. In: Proceedings of the Eleventh ACM International Conference on Web Search and Data Mining, pp. 583–591 (2018)
9. Mikolov, T., Chen, K., Corrado, G., Dean, J.: Efficient estimation of word representations in vector space. arXiv preprint arXiv:1301.3781 (2013)
10. Wang, H., Wang, J., Wang, J., Zhao, M., et al.: GraphGAN: graph representation learning with generative adversarial nets. arXiv preprint arXiv:1711.08267 (2017)
11. Hofmann, T., Buhmann, J.: Multidimensional scaling and data clustering. In: Advances in Neural Information Processing Systems, pp. 459–466 (1995)
12. Balasubramanian, M., Schwartz, E.L.: The Isomap algorithm and topological stability. Science **295**(5552), 7 (2002)
13. Roweis, S.T., Saul, L.K.: Nonlinear dimensionality reduction by locally linear embedding. Science **290**(5500), 2323–2326 (2000)
14. Belkin, M., Niyogi, P.: Laplacian eigenmaps and spectral techniques for embedding and clustering. Adv. Neural. Inf. Process. Syst. **14**(6), 585–591 (2001)
15. Zhang, D., Yin, J., Zhu, X., Zhang, C.: Network representation learning: a survey. arXiv preprint arXiv:1801.05852 (2017)
16. Krioukov, D., Papadopoulos, F., Kitsak, M., Vahdat, A., Boguná, M.: Hyperbolic geometry of complex networks. Phys. Rev. E **82**(3), 036106 (2010)
17. Goodfellow, I., Pouget-Abadie, J., Mirza, M., Xu, B., et al.: Generative adversarial nets. In: Advances in Neural Information Processing Systems, pp. 2672–2680 (2014)
18. Gulrajani, I., Ahmed, F., Arjovsky, M., et al.: Improved training of wasserstein gans. In: Advances in Neural Information Processing Systems, pp. 5769–5779 (2017)
19. Mikolov, T., Sutskever, I., Chen, K., Corrado, G.S., Dean, J.: Distributed representations of words and phrases and their compositionality. In: Advances in Neural Information Processing Systems, pp. 3111–3119 (2013)
20. McCallum, A.K., Nigam, K., Rennie, J., Seymore, K.: Automating the construction of internet portals with machine learning. Inf. Retr. **3**(2), 127–163 (2000)

21. Sen, P., Namata, G., Bilgic, M., Getoor, L., Galligher, B., Eliassi-Rad, T.: Collective classification in network data. AI Mag. **29**(3), 93 (2008)
22. Tang, L., Liu, H.: Relational learning via latent social dimensions. In: Proceedings of the 15th ACM SIGKDD International Conference on Knowledge Discovery and Data Mining, pp. 817–826 (2009)
23. Tang, L., Liu, H.: Leveraging social media networks for classification. Data Min. Knowl. Discov. **23**(3), 447–478 (2011)
24. Fan, R.E., Chang, K.W., Hsieh, C.J., et al.: LIBLINEAR: a library for large linear classification. J. Mach. Learn. Res. **9**(Aug), 1871–1874 (2008)
25. Liben-Nowell, D., Kleinberg, J.: The link-prediction problem for social networks. J. Assoc. Inf. Sci. Technol. **58**(7), 1019–1031 (2007)
26. Maaten, L.V.D., Hinton, G.: Visualizing data using t-SNE. J. Mach. Learn. Res. **9**(Nov), 2579–2605 (2008)

Beyond Pivot for Extracting Chinese Paraphrases

Yu Zhang$^{(\boxtimes)}$, Le Qi, Linjie Wang, Linlin Yu, and Ting Liu$^{(\boxtimes)}$

Research Center for Social Computing and Information Retrieval,
Harbin Institute of Technology, Harbin, China
{zhangyu,lqi,linjiewang,llyu,tliu}@ir.hit.edu.cn

Abstract. Paraphrasing is a critical issue in many Natural Language Processing (NLP) applications. The traditional Pivot-based methods of extracting paraphrases require a large-scale bilingual parallel corpus. The quality of the extracted paraphrases is affected by the quality of bilingual parallel corpora and word alignment. In this paper, we propose a method for Chinese paraphrases extraction. An online translation system is used to obtain the candidate paraphrases of a word. A deep neural network model combined with cosine similarity is exploited to filter the candidate results through computing the similarity of word vectors between a word and its candidate paraphrase. Experiments are conducted in two ways: (1) The random sampling is employed to manually verify the correctness of the paraphrases results. The effect has been significantly improved; (2) We design two Question Answering (QA) systems based on the NLPCC2016 Document Based Question Answering (DBQA) corpus. One uses the BM25 model to retrieve the candidate answer sentences, and another uses the Convolution Neural Network (CNN) model. Extracted paraphrases are quite effective in question reformulation, enhancing the MRR from 56.33% to 60.21% (BM25) and from 63.82% to 66.60% (CNN) with the questions of NLPCC 2016 DBQA corpus.

Keywords: Paraphrase extraction · Pivot · Question Answering
Neural network

1 Introduction

Paraphrases are alternative ways to convey the same information [4]. Paraphrasing has always been a very important issue since it is significant to many NLP applications, including Machine Translation (MT), Information Retrieval (IR), Summarization, Question Answering (QA), and Natural Language Generation (NLG) [16]. Especially for a QA system, if an input question can be expanded with its various paraphrases, the recall of answers can be improved [17].

There are many paraphrase extraction methods, such as paraphrasing using WordNet [6], patterns [3] and "pivot" method [1]. Paraphrasing with Pivot is widely used because it is efficient and easy to obtain the corpus. In traditional

© Springer Nature Switzerland AG 2018
S. Zhang et al. (Eds.): CCIR 2018, LNCS 11168, pp. 132–144, 2018.
https://doi.org/10.1007/978-3-030-01012-6_11

"pivot" methods, word alignment needs to be done between the bilingual parallel sentences. The word alignment error may lead to a large number of mistakes during the paraphrases extraction. Furthermore, polysemy is a pervasive phenomenon in human language. Thus, even two Chinese words have the same translation of foreign language, the meaning of the two words may differ. For example, " 法庭 " (court) and " 网球场 " (tennis court) has the same translation "court".

In this paper, to avoid the wrong translation caused by the alignment error, an online translation (OT) system is used to acquire candidate paraphrases. Word alignment is no longer need to do in the bilingual parallel corpora. A Feedforward Neural Network (FNN) combined with the cosine similarity between word vectors is exploited to filter the candidate paraphrases.

The remainder of the paper is organized as follow: our method of extracting Chinese paraphrases is presented in Sect. 2; Experiments results and analysis are discussed in Sect. 3; Section 4 introduces related work; and the Sect. 5 is the conclusion and future work.

2 Our Method

In this section, we will describe in details the algorithm used to extract the paraphrase of Chinese words using the English translation Tools. As shown in Fig. 1, we first acquire the English translation of Chinese words, then use the translation resources to extract paraphrases. After that, we calculate a score for each pair of candidate paraphrases and rank the candidate paraphrases by it. Meanwhile, we set a threshold α to filter candidate paraphrases.

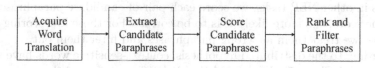

Fig. 1. The flow chart of the paraphrase extraction algorithm.

2.1 Word Translation Acquisition

In order to avoid alignment errors leading to get wrong translations of the original word, we use OT to acquire the translations. The OT provides a wealth of dictionary resources and the translation results are edited manually with high accuracy. In addition, OT resources also include idioms on the Internet without the limitations of traditional dictionary.

2.2 Candidate Paraphrases Extraction

In traditional pivot method, the vocabulary of the target language in the bilingual parallel corpus is taken as a pivot, and all the source language words aligned with the pivot are taken as candidate paraphrases. We now extract the paraphrases directly from the English translations of the Chinese words obtained in Sect. 2.1. For the original word w in the source language, we treat all words with the same English translation T as the candidate paraphrases C of w. For example, the Chinese words: " 母亲 " and " 妈妈 " have the same English translation "mother", so "mother" is the pivot which connects " 妈妈 " and " 母亲 ". As a result, we will extract " 母亲 " as a candidate paraphrase of " 妈妈 ".

By the above method, we extract paraphrases on 183,728 words, and get 2,739,270 candidate paraphrase pairs. The number of candidate paraphrase pairs we extract are much larger than that obtained by thesaurus.

2.3   Candidate Paraphrases Filter and Ranking

In the candidate paraphrase pairs extracted by the method described in Sect. 2.2, some are correct, such as " 辞职信 " and " 辞呈 " with the translation as "resignation". But some are wrong, such as " 清洁度 (the degree of cleanliness)" and " 干净 (cleanliness)" where the pivot is "cleanliness". So not all of words with the same translation are paraphrases. After analyzing the paraphrase results, we find that the main reason for the mistake is that a foreign language words may have multiple meanings rather than a single meaning. For example, one of the Chinese meanings of "watch" is " 收看 ", meaning that "to look at somebody/something for a time" and another one is " 手表 ", meaning that "a type of small clock that you wear on your wrist". Obviously " 收看 " and " 手表 " are not paraphrase, but are extracted by "watch". The polysemy of the word is the most important reason leading to extract wrong paraphrases.

In order to solve this problem and facilitate the use of lexical paraphrase resources in other NLP tasks, we score each pair of candidate paraphrases. The higher the score, the more likely it is to be correct. For those low-scoring candidate pairs, we treat them as wrong paraphrases and filter them out.

According to the distributed hypothesis, context-sensitive words have similar semantics, and Chan [7] also proved that it is effective to use the context similarity to sort the obtained paraphrase resources. Meanwhile, the word vector is the representation of the semantic meaning of the words in the continuous space. The similarity of word vectors can indicate the correlation between the words. At present, the common methods of learning the word vectors are all based on the distributed hypothesis. Therefore, we use word vectors as the input to our model to exploit contextual and semantic information. In addition, the pivot is also important to measure the correlation between the word and its paraphrase. Therefore, we incorporate the number of co-owned translations of the word and its paraphrase as a feature into the model.

Based on the above discussion, we employ a FNN to model original word w and one word c_i of its candidate paraphrases C. The network makes use of the information of word vector and co-owned translation, learns the correlation between words and score the candidate paraphrase c_i. Figure 2 shows the architecture of the model.

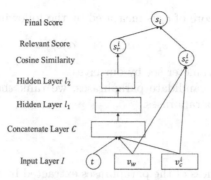

Fig. 2. Model architecture.

The inputs to the model are the word vectors of w and c_i, and the number of the same translations of two words. We concatenate all the inputs together as the feature vector that contains the information of w and c_i. Then the feature vector will be passed through two hidden layers in the network, and mapped into a low-dimensional feature space. The resulting vector can be viewed as a low-dimensional feature embedding including the relevant information between w and c_i. Next, the model computes a similarity score between w and c_i based on the low-dimensional feature embedding. The score is normalized into probability using a *Sigmoid* function. In addition, we compute cosine similarity between the word vectors of w and c_i as another score. Finally, the weighted sum of the two scores is viewed as the final score of c_i.

Formally, let v_w and v_c^i be the vectors of embedding representation of w and c_i respectively, and let t be the number of co-owned translations of w and c_i. In addition, let C be the feature vector, feature represent the low-dimensional feature vector, l_1 is the intermediate hidden layers, W_i is the weight matrices associated with x in hidden layer i, and b_i denotes the bias terms associated with x. We then have:

$$C = v_w \oplus v_c^i \oplus t \qquad (1)$$
$$l_1 = f(W_1 C + b_1) \qquad (2)$$
$$l_2 = f(W_2 l_1 + b_2) \qquad (3)$$
$$feature = l_2 \qquad (4)$$

where f (*tanh* in our model) is the activation function at hidden layers.

The relevance score s_r between w and c_i is calculated and normalized by the *sigmoid* function with the weight matric W_s and bias term b_s:

$$s_r^i = sigmoid(W_s y + b_s) \qquad (5)$$

In addition, the cosine similarity between w and c_i is calculated as follow:

$$s_c^i = cosine(v_w, v_c^i) \qquad (6)$$

At last, the final score of c_i is measured by the following equation:

$$s_i = \alpha s_r^i + (1 - \alpha)s_c^i \tag{7}$$

where α is a hyper parameter set by ourselves.

After scoring each candidate paraphrases, we rank them by the score and filter out low-scoring paraphrases.

3 Experiments

To verify the effectiveness of the paraphrases extracted by the method proposed in this paper, we designed two experiments. Since there is no standard test set for paraphrases, the first experiment is set to evaluate the extracted paraphrases manually. Another experiment is designed to verify whether it is helpful for QA Systems. We designed two QA systems, used BM25 and CNN models respectively for extracting the candidate answer sentences. Extracted paraphrases are used in question formulation.

Our model for paraphrase filtering contains 3 parameters. The dimensions of the two hidden layers are 100 and 50 respectively. The alpha value is set as 0.3 due to our attempts on the development set. At last, the 100-dimensional word embedding is pre-trained by Word2Vec [13] algorithm with Sogou corpus.

3.1 Manual Evaluation on the Extracted Paraphrases

We used the API[1] provided by Youdao for OT to obtain the English translation of Chinese words. The basic and the online interpretations of the vocabulary returned by the API are treated as the foreign translation. Sogou Internet news data set[2] (2012 edition) was used as monolingual corpus, and the LTP (Language Technology Platform, provided by Harbin Institute of Technology) was used to preprocess the corpus, such as word segmentation. The word vectors were trained by the processed corpus, and the vocabulary was also obtained from processed corpus. We used the API to obtain English translation of Chinese words, and got 583,348 English translation words.

Because we do not know how many paraphrases are there for a specific word, so it is difficult to compute the *recall* value. In our experiment, *Precision* and *MRR* are used to evaluate the results. *Precision* is used to measure the proportion of the correct results in extracted paraphrase results. *MRR* (8) is used to evaluate the ranking for the candidate paraphrases.

$$MRR = \frac{1}{|P|} \sum_{i=1}^{|P|} \frac{1}{rank_i} \tag{8}$$

where $|P|$ is the number of words, $rank_i$ is the position of the first correct result for the $i-$th word.

---

[1] http://fanyi.youdao.com/openapi.do.
[2] http://www.sougou.com/labs.

80,553 individual words were obtained from the monolingual corpus, and paraphrases were extracted for these words. The Table 1 reports some examples of paraphrases.

Table 1. Some examples of paraphrases.

Words	Paraphrases	Words	Paraphrases
脸面 (face)	面子 颜面	急挫 (plunge)	狂跌 大跌
从军 (join the army)	当兵 参军 服役 入伍	牲畜 (livestock)	畜禽 牲口 家畜
声势 (momentum)	气势	年代久远 (age-old)	古老 年深日久
谋杀案 (murder case)	杀人案 凶杀案	禁不住 (be unable to bear)	情不自禁 不禁
少许 (a little)	少量 一点点	此次 (this time)	本次 这次

In this experiment, we randomly selected 500 individual Chinese words and extracted their paraphrases by different methods. The selected words and their paraphrases are annotated manually. We have 5 annotators. For each pair of paraphrases, we vote on the annotator's label and follow the principle of minority obeying majority. The methods we chose in the experiment are shown as follow:

PPDB (the ParaPhrase DataBase): Paraphrases in PPDB are extracted from the Bilingual Parallel Corpus. The extracted paraphrases are divided into different sizess. The smaller the vocabulary is, the higher the quality is. [8]

PPDB + OT (Online Translation) Filter: In this method, we use the online translation API to filter the paraphrases in PPDB.

PPDB + Re-ranking: The candidate paraphrases for each word are re-ranked by our method.

Pivot: We use the OT API to extract paraphrases for each word.

Pivot + Cosine: We re-rank and filter out the candidate paraphrase pairs by the cosine similarity between the word vectors of them.

Pivot + Re-ranking: This is the method proposed in this paper.

Table 2. The compared results for paraphrases.

Methods	Precision	MRR
PPDB	36.60%	70.70%
PPDB + OT Filter	48.20%	77.00%
PPDB + Re-ranking	36.60%	72.50%
Pivot	13.20%	10.90%
Pivot + Cosine	66.80%	90.10%
Pivot + Re-ranking	81.70%	94.80%

Table 2 reports the evaluation results for paraphrases. The method using "Pivot" reports the worst performance, and the phenomenon of polysemy is the main problem resulting in a large number of error results. The *precision* of "PPDB" is only 36.6%, because of errors introduced in bilingual sentences alignment. After filtering by the OT API, the effect has some promotion that can be seen in the results of "PPDB + OT Filter". By using re-ranking in "PPDB", the MRR score has been improved. It proves that the score calculated by our re-ranking method is superior to the score calculated in "PPDB".

In "Pivot + Cosine", candidate paraphrases of a word are re-ranked according to the similarity between the word and its paraphrases. The closer the word vectors are, the more similar the word meaning is. It can filter out false paraphrases introduced by the "Pivot". Our method, the "Pivot + Re-ranking", has achieved the best results. The similarity score is calculated by cosine similarity combined with a FNN. The semantic relationship between words can be learned by the FNN model to improve the quality of the paraphrases.

3.2   Evaluate the Extracted Paraphrases by QA Systems

In this part, the experiments aim to verify the role of paraphrase in QA systems. The dataset of NLPCC (Conference on Natural Language Processing and Chinese Computing) evaluation on DBQA (Document Based Question Answering system) is used in these experiments. There are 181,882 question-answer pairs in the training set, and 122,531 question-answer pairs in the test set. Since more than 20 candidate answers corresponds to the same question, so there are nearly 8,000 questions in the training set and 6,000 questions in the test set.

MRR and MAP are used to evaluate the results in these experiments. MRR is the same as the Eq. (10), but $|P|$ here stands for the number of questions. MAP is the average precision, shown in Eq. (11).

$$MAP = \frac{1}{|Q|} * \sum_{i=1}^{|Q|} AveP(C_i, A_i) \tag{9}$$

$$AveP(C, A) = \frac{\sum_{k=1}^{n} (p(k) * rel(k))}{min(m, n)} \tag{10}$$

where $|Q|$ is the number of questions. k is the ranking of the retrieved answer sentences. m is the number of the correct answer sentences. n is the number of all the candidate answer sentences. $p(k)$ is the accuracy of the previous k sentences. $rel(k)$ is set to "1" if the kth sentence is the correct answer, otherwise it is set to "0".

In this part, BM25 and CNN models are used to extract the answer sentences. BM25 is a kind of word matching method introduced by IR, so it will be more dependent on paraphrasing. Although, the input of the CNN model introduces semantic information, we want to know whether it plays a role for QA using paraphrase.

Some compared experiments have done based on the BM25 method.

BM25: BM25 is used to calculate the relevance between the question and the candidate answer sentences. The score is calculated by the Eq. (13).

$$Score(Q, d) = \sum_{i}^{n} W_i * R(q_i, d) \tag{11}$$

$$R(q_i, d) = \frac{f_i * (k_1 + 1)}{f_i + K} * \frac{qf_i * (k_2 + 1)}{qf_i + k_2} \tag{12}$$

$$K = k_1 * (1 - b + b * \frac{dl}{avgdl}) \tag{13}$$

where $R(q_i, d)$ is the relevance between words and document. W_i is the weight for each word. k_1, k_2 and b are the regulatory factors, which $k_1 = 2$, $k_2 = 1$ and $b = 0.75$ based on empirical value. f_i is the term frequency of a word in the document. qf_i is the term frequency of a word in the query. dl is the length of the document. $avgdl$ is the average length of the documents.

BM25 + Sim-WordVectors: In this method, word vectors are used to calculate the similarity between the question and the candidate answer sentence.

BM25 + HowNet: HowNet[3] is used to extract the candidate answer sentences.

BM25 + Paraphrasing: Extracted paraphrases by the method proposed in this paper are used in the question formulation. We treat a word and its paraphrases as the same word.

Table 3. The compared results for QA systems based on BM25.

Methods	MRR	MAP
BM25	56.33%	56.23%
BM25 + Sim-WordVectors	57.88%	57.75%
BM25 + HowNet	55.91%	55.86%
BM25 + Paraphrasing	60.21%	60.15%

The experiment results are shown in Table 3. When using paraphrases for question formulation, the effect is approximately 4% higher than the BM25 baseline. Since the keyword matching is very important in BM25 to find the candidate answering sentences, so it is more likely to find the answer if the paraphrases for the keyword are used in question. In "BM25 + Sim-WordVectors", word vectors are used to calculate the similarity between the question and the candidate answer sentence. Some sentences maybe more similar to the question even though their keywords not exactly match. But similar word vectors not always have the same meaning, such as "good" and "bad", some mistakes maybe introduced in this method. In "BM25 + HowNet", we use the Chinese

---

[3] http://www.keenage.com/.

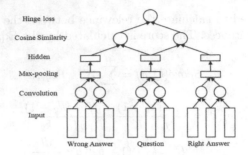

Fig. 3. Model architecture.

semantic dictionary for extracting candidate answer sentences. The results are not good as expected. With HowNet, we can find more relative keywords, but maybe they are not paraphrases. Resulting in worse results in the performance of this method.

Another experiments for candidate answer sentences selection has done based on the CNN model, similar with the work of Tan [15]. The main purpose is to use the convolution neural network to learn the vector representation of a sentence. The architecture of the CNN model for candidate answer sentences selection is illustrated in Fig. 3.

As it can be seen from the Fig. 3, the model has three sentences as input: questions, correct answers and wrong answers. The main idea is to convolute three sentences with the convolution layer and the pooling layer. The corresponding sentence vector representation will be got through learning. Then, a hidden layer changes the vector dimension. At last, cosine similarity is used to calculate the similarity. "Hinge loss" is used as the cost function rather than the conventional cross entropy, in order to make the correct answer be able to score ahead of the other sentences, not just close to 1.

Convolution layer: Use the convolution kernel W_c sliding window to extract the features of the input text X, and finally get the context feature vector c_t (14).

Pooling Layer: Extract the global information in the sentence level using the output vector of the convolution layer. The MaxPooling is used to collect the global information v_t of c_t (15).

Hidden Layer: Use a fully connected layer to change the dimension of a sentence vector (16).

Output Layer: Firstly, similarity between the question and the correct answer sentence, the question and the wrong answer sentence are calculated by cosine similarity (17). Then calculate the cost of the whole training sample (18).

$$c_t = tanh(W_c * X + b_c) \tag{14}$$
$$v_t = max_{t=1,\cdots,T} c_t(i) \tag{15}$$

$$h_t = \sigma(W_h * v_t + b_h) \tag{16}$$

$$s_i = cosine(q, a) \tag{17}$$

$$cost = max(t_0 - s_1 + s_2, 0) \tag{18}$$

where we set t_0 as 0.5.

During test, we only use the half of the model in Fig. 3 (the left or right half without Hinge Loss layer). We just calculate the similarity between each pair of questions and the candidate answers as the score for each candidate answer, and then sort the candidate answers based on the score. Finally, we choose the candidate with the highest score as the correct answer to the question.

If we use paraphrase in the model, we replace the embedding of a word with the weighted average of the word embedding and its paraphrases' embedding.

Table 4. The compared results for QA systems based on CNN Model.

Methods	MRR	MAP
CNN (QA Corpus)	53.63%	53.48%
CNN (BAIKE Corpus)	63.82%	63.63%
CNN (BAIKE Corpus) + Paraphrasing	66.60%	66.35%

The experimental results are reported in Table 4. The word embeddings are trained by using different kinds of corpus. CNN (QA corpus) is the model trained by the QA evaluation corpus provided by NLPCC, and CNN (BAIKE corpus) the model trained by the BAIKE corpus download from the http://baike.baidu. com (more than 4,400,000 of words and content for each word are in include in this corpus). The result of CNN (BAIKE corpus) is better than CNN (QA corpus). Thus, paraphrases are used in question formulation based on CNN (BAIKE corpus). The MRR rose from 63.82% to 66.60%. So we can see that paraphrasing also work well with deep learning model. By using paraphrases in question formulation, we can get a candidate answer sentence that is more similar to the meaning of the question, not just to word vectors.

4   Related Work

Different algorithms for extracting paraphrases use different types of corpus resources, such as thesaurus, monolingual corpus, monolingual parallel corpora, monolingual comparable corpora and bilingual parallel corpus.

In the thesaurus resources such as WordNet and HowNet, the words are annotated according to the semantic relation of words, so semantic related words can be found according to the corresponding annotation information. Synonyms in the thesaurus are treated as lexical paraphrases [11]. The hyponymy is also used to extract paraphrases [2]. Lexical paraphrases obtained using thesaurus resource are of very good quality. Thesaurus is always labeled by hand, where the number of words is small, the resulting number of paraphrases is limited.

The monolingual materials are very rich, and news from the Internet can be used as a monolingual corpus. The method of extracting paraphrases from the monolingual corpus is mainly based on the distributed hypothesis [5,12]. The monolingual materials are easy to get, and the size of the corpus is large. The paraphrases extracted by this method are context similar, but the meaning of the words is not necessarily the same.

A monolingual parallel corpus is a text that describes the same content in the same language. Barzilay et al. [4] used different English translation versions of the same literary work as a monolingual parallel Corpus and extracted paraphrases from it. Extract paraphrases from the monolingual parallel corpus needs to align the sentences in the corpus, and then use the template to extract the paraphrases from it. The accuracy of this method is high, but the monolingual parallel corpora are few, which restricts the scale of the results.

A monolingual comparable corpus is a related text that describes the same subject or event [10]. Nakagawa et al. [14] extract paraphrases from a monolingual comparable corpus that describe the news of the same event. It's easier to get monolingual comparable news, but there are not necessarily paraphrased sentences in the news. It is difficult to extract paraphrases from monolingual comparable corpus, and the quality of extracted paraphrases is not high.

A bilingual parallel corpus is a textual corpus of two different languages that describe the same information. The paraphrase extraction method from the bilingual parallel corpus is mainly based on the "pivot" method [1]. In this method, we consider that the meanings of the words with the same foreign translation are the same. The corpus used in this method is in large scale and not limited to specific fields. Since this method needs to align sentences in Bilingual Parallel Corpus, alignment errors lead to a number of errors in the final extraction of the paraphrases. For example, in PPDB [9], " 开心 "(happy) and " 亲属 "(relatives) are treated as paraphrases. It is due to the alignment error. Polysemy can also cause mistakes in this method.

5   Conclusion and Future Work

In this paper, an OT system is used to get the candidate paraphrases of a word. A deep neural network model combined with cosine similarity is exploited to filter the candidate results through computing the similarity of word vectors between paraphrases of a word. The experiment results showed that the extracted paraphrases through this method have high quality, and are also helpful for NLP applications such as QA systems. In the future, our research work focuses on two aspects. One is to determine the probability of the replaceable of a word as the paraphrases. Since not all the paraphrases of a word can replace each other, so the confidence degree can help users to determine whether a word can be replaced by another word or not. Another more meaningful research work will be to extract phrase-level paraphrases from large-scale corpus.

Acknowledgments. This work was supported by the National Basic Research Program (973 Program) (Grant No. 2014CB340503) and the National Natural Science Foundation of China (Grant No. 61472105 and 61502120).

References

1. Bannard, C., Callison-Burch, C.: Paraphrasing with bilingual parallel corpora. In: Proceedings of the 43rd Annual Meeting on Association for Computational Linguistics, pp. 597–604. Association for Computational Linguistics (2005)
2. Barzilay, R., Elhadad, M.: Using lexical chains for text summarization. In: Advances in Automatic Text Summarization, pp. 111–121 (1999)
3. Barzilay, R., Lee, L.: Learning to paraphrase: an unsupervised approach using multiple-sequence alignment. In: Proceedings of the 2003 Conference of the North American Chapter of the Association for Computational Linguistics on Human Language Technology, vol. 1, pp. 16–23. Association for Computational Linguistics (2003)
4. Barzilay, R., McKeown, K.R.: Extracting paraphrases from a parallel corpus. In: Proceedings of the 39th Annual Meeting on Association for Computational Linguistics, ACL 2001, pp. 50–57. Association for Computational Linguistics, Stroudsburg (2001). https://doi.org/10.3115/1073012.1073020
5. Bhagat, R., Ravichandran, D.: Large scale acquisition of paraphrases for learning surface patterns. In: Proceedings of ACL-08: HLT, pp. 674–682 (2008)
6. Bolshakov, I.A., Gelbukh, A.: Synonymous paraphrasing using wordnet and internet. In: Meziane, F., Métais, E. (eds.) NLDB 2004. LNCS, vol. 3136, pp. 312–323. Springer, Heidelberg (2004). https://doi.org/10.1007/978-3-540-27779-8_27
7. Chan, T.P., Callison-Burch, C., Van Durme, B.: Reranking bilingually extracted paraphrases using monolingual distributional similarity. In: Proceedings of the GEMS 2011 Workshop on GEometrical Models of Natural Language Semantics, pp. 33–42. Association for Computational Linguistics (2011)
8. Ganitkevitch, J., Callison-Burch, C.: The multilingual paraphrase database. In: LREC, pp. 4276–4283 (2014)
9. Ganitkevitch, J., Van Durme, B., Callison-Burch, C.: PPDB: the paraphrase database. In: Proceedings of the 2013 Conference of the North American Chapter of the Association for Computational Linguistics: Human Language Technologies, pp. 758–764 (2013)
10. Ho, C., Murad, M.A.A., Doraisamy, S., Kadir, R.A.: Extracting lexical and phrasal paraphrases: a review of the literature. Artif. Intell. Rev. **42**(4), 851–894 (2014)
11. Kauchak, D., Barzilay, R.: Paraphrasing for automatic evaluation. In: Proceedings of the main conference on Human Language Technology Conference of the North American Chapter of the Association of Computational Linguistics, pp. 455–462. Association for Computational Linguistics (2006)
12. Lin, D., Pantel, P.: DIRT@ SBT@ discovery of inference rules from text. In: Proceedings of the Seventh ACM SIGKDD International Conference on Knowledge Discovery and Data Mining, pp. 323–328. ACM (2001)
13. Mikolov, T., Sutskever, I., Chen, K., Corrado, G.S., Dean, J.: Distributed representations of words and phrases and their compositionality. In: Advances in Neural Information Processing Systems, pp. 3111–3119 (2013)

14. Nakagawa, H., Masuda, H.: Extracting paraphrases of Japanese action word of sentence ending part from web and mobile news articles. In: Myaeng, S.H., Zhou, M., Wong, K.-F., Zhang, H.-J. (eds.) AIRS 2004. LNCS, vol. 3411, pp. 94–105. Springer, Heidelberg (2005). https://doi.org/10.1007/978-3-540-31871-2_9
15. Tan, M., Santos, C.d., Xiang, B., Zhou, B.: LSTM-based deep learning models for non-factoid answer selection. arXiv preprint arXiv:1511.04108 (2015)
16. Zhao, S., Wang, H., Liu, T.: Paraphrasing with search engine query logs. In: Proceedings of the 23rd International Conference on Computational Linguistics, pp. 1317–1325. Association for Computational Linguistics (2010)
17. Zhao, S., Zhou, M., Liu, T.: Learning question paraphrases for QA from Encarta logs. In: IJCAI, pp. 1795–1801 (2007)

Information Extraction and Sentiment Analysis

Joint Attention LSTM Network
for Aspect-Level Sentiment Analysis

Guoyong Cai[1,2(✉)] and Hongyu Li[1(✉)]

[1] Guilin University of Electronic Technology, Guilin, China
ccgycai@gmail.com, HongyuLi009@163.com
[2] Guangxi Key Laboratory of Trusted Software, Guangxi, China

Abstract. Aspect-level sentiment analysis, as an important type of sentiment analysis, is a fine-grained sentiment analysis task which has received much attention recently. Recent work combines attention mechanisms with neural networks to learn aspects feature and achieves state-of-the-art performance. However, the prior work ignores the sentiment terms feature and the latent correlation between aspect terms and sentiment terms. In order to make use of aspects terms and sentiment terms information, a method that based on joint attention LSTM network (JAT-LSTM) for aspect-level sentiment analysis is proposed, which aspect attention and sentiment attention are combined to construct a joint attention LSTM network. The experimental results on the benchmark datasets show that the proposed method achieves better performance than the current state-of-the-art.

Keywords: Sentiment analysis · Aspect-level · Attention mechanism
LSTM · Deep learning

1 Introduction

Sentiment analysis, also known as opinion mining, is a field of study that analyzes the sentiments, attitudes and opinions contained in user generated content [1]. Aspect-level sentiment analysis, as an important type of sentiment analysis, is a fine-grained sentiment analysis task, it aims to determine the sentiment polarity of specific aspects [2]. Take the review "The food is great, but the service is terrible" for an example, people expressed opposite opinions on aspect "food" and "service" of the entity "restaurant". Mining the sentiment of reviews at aspect-level can help businesses to discover the advantages and disadvantages of products and make better product design, and it can also help consumers understand the comprehensive evaluation of various aspects of a product and make better purchase decisions.

In recent years, neural networks have made significant progress, along with its successful applications in many NLP tasks [3–6], more and more researchers tend to use deep learning methods for aspect-level sentiment analysis. Sentiment classifier constructed with deep learning methods can automatically learn a large number of hidden features from the original datasets through its deep, nonlinear network structure, obtaining a higher level and more abstract representation, and can achieve excellent results in sentiment classification.

© Springer Nature Switzerland AG 2018
S. Zhang et al. (Eds.): CCIR 2018, LNCS 11168, pp. 147–157, 2018.
https://doi.org/10.1007/978-3-030-01012-6_12

Attention mechanism has been widely used in various fields of deep learning in recent years, especially in the tasks of image recognition [7], machine translation [8], and sentence classification [9], etc. The neural networks combine with attention mechanism have achieved great successes in various tasks. The attention-based neural network can make the neural network pay more attention to the important information for accomplishing the current task in the training process, and capture more detailed feature information of the target.

Motivated by the above intuition, we propose an aspect-level sentiment analysis model based on the joint attention LSTM (Long Short Term Memory) network, which combines the attention mechanism to construct a joint attention LSTM network and apply it to the aspect-level sentiment analysis task. Finally, the experiments are performed on two benchmark datasets, and the results show that the proposed method can achieve better performances on aspect-level sentiment classification.

2   Related Work

Traditional approaches for aspect-level sentiment analysis are to design a set of sentiment features, and using the sentiment feature information of texts to identify the sentiment polarity of specific aspects. Hu and Liu [10] used the adjectives extracted from texts as sentiment terms, and resorted to SentiWordNet to analyze the orientations of all the adjectives, then predicted the sentiment polarity of different aspects of a product. Yu et al. [11] utilized positive and negative reviews to train a one-class SVM (Support Vector Machine) to identify aspects, and train a SVM sentiment classifier to determine the sentiment polarity of aspects. These studies have been proved to be quite effective for determining the sentiment polarity of aspects. However, these dictionary-based or traditional machine learning methods require artificially construct sentiment lexicons and design a set of features, leading to a great application limitations.

With the great development of deep learning, more and more researchers tend to apply neural networks to deal with aspect-level sentiment analysis task. Nguyen et al. [12] proposed an extended model of RNN (Recurrent Neural Network) to make the representation of the target aspect richer by using syntactic information from both the dependency and constituent trees of the sentence. Tang et al. [13] proposed two target-dependent LSTM models by incorporating the target aspect information, and have achieved a good performance for aspect-level sentiment classification. Lakkaraju et al. [14] proposed a deep learning based framework to joint modeling the aspects and sentiment without making strict modeling assumptions about the interleaving of aspects and sentiment extraction phases, the experiment results show that the model achieved better performance for aspect-level sentiment classification.

The above studies utilize deep learning methods to better learn the features information from text, and achieve better performance on aspect-level sentiment classification than traditional methods. However, these methods assume that the sentence contains only one object or one specific aspect while the reviews in real life often contain different opinions toward multiple aspects of products or services, hence it is difficult to extract different features for different aspects from the same review. To solve this problem, researchers design attention-based neural networks to deal with aspect-level sentiment classification task. Using attention mechanism can help neural networks to focus on the important aspects information in the training process, and learn different features for different aspects. Liang et al. [15] proposed a multi-attention convolutional neural network, the networks can capture deeper level sentiment information and distinguish sentiment polarity of different aspects through a multi-attention mechanism. Wang et al. [16] proposed an attention-based LSTM model, by taking the given aspect terms as attention information to embed into the input and output of the LSTM network, enforcing the model to attend to the features of different aspects in the training process, then identified the sentiment polarity of different aspects effectively. Cheng et al. [17] proposed a hierarchical attention network, using an aspect attention mechanism to extract aspect-related information which can guide the neural network to better capture the sentiment feature of texts, and achieved the state-of-the-art performance for aspect-level sentiment classification.

The prior works only attend to the given aspect terms information and ignore the sentiment terms information corresponding to the aspect terms of texts. In this paper, we consider the important influence of aspect terms information and sentiment terms information on the representation of a sentence, combining an attention mechanism to construct a joint attention LSTM network (JAT-LSTM) for aspect-level sentiment analysis task to improve the aspect-level sentiment classification accuracy.

3   Joint Attention LSTM Network

This section presents the proposed joint attention LSTM network (JAT-LSTM) model for aspect-level sentiment analysis. The architecture of JAT-LSTM model is shown in Fig. 1, which includes three modules: (1) input module (as shown in (A) of Fig. 1), (2) joint attention module (as shown in (B) of Fig. 1), (3) sentiment classification module (as shown in (C) of Fig. 1). In the input module, the model firstly combines the aspect term embedding and sentiment term embedding with sentences embedding respectively, and learns the contextual features of each sentence through the corresponding LSTM network. Then, in the joint attention module, the model capture aspect feature and sentiment feature with aspect attention and sentiment attention respectively, and obtain the sentence feature representations which contain important aspect information and sentiment information. Finally, the sentiment classification module output the probability of classifying the sentiment polarity of aspects. The details of the model are described in the Sects. 3.1, 3.2 and 3.3.

3.1   Input Module

The input of the proposed model contains sentences embedding X, aspect term embedding v_a and sentiment term embedding v_s. In order to make better use of the given aspect terms information and corresponding sentiment terms information in the sentence, we use two different LSTM networks to learn the contextual feature of each word in the input module: $LSTM^{(a)}$ network is used to learn the interdependence information between the given aspect terms and each word in the sentence (as shown in (A-1) of

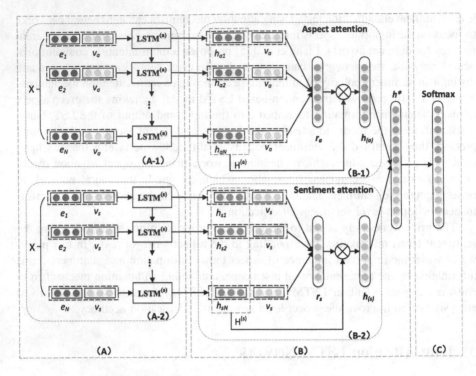

Fig. 1. The architecture of joint attention LSTM network.

Fig. 1), $LSTM^{(s)}$ network is used to learn the interdependence information between the given sentiment terms and each word in the sentence (as shown in (A-2) of Fig. 1).

In order to get a higher quality word embedding, we first train word embedding on an external large domain-specific corpus, and obtain a word embedding matrix $\mathbb{E} \in \mathbb{R}^{d_w \times |\mathcal{W}|}$, where d_w is the dimension of word embedding and $|\mathcal{W}|$ is the size of vocabulary. For the sentences embedding X, we treat each sentence as a sequence of words in word units, then we use the word embedding matrix $\mathbb{E}$ to project each word in the sentence into a word embedding, and obtain the sentences embedding $X = [e_1, e_2, \ldots e_i, \ldots e_N]$, where $e_i \in \mathbb{R}^{d_w}$ is the word embedding of word i, N represents the length of the sentence. Similarly, we use the word embedding matrix $\mathbb{E}$ to project all the given aspect terms and sentiment terms into aspect embedding $\Phi_a \in \mathbb{R}^{d_w \times |\mathcal{A}|}$ and sentiment embedding $\Phi_s \in \mathbb{R}^{d_w \times |\mathcal{S}|}$ respectively, where $|\mathcal{A}|$ and $|\mathcal{S}|$ represent the number of all given aspect terms and sentiment terms respectively, we use v_a to represent aspect term embedding and v_s to represent sentiment term embedding.

The input of $LSTM^{(a)}$ network contains sentences embedding X and aspect term embedding v_a. We first extend the given aspect term embedding to an aspect term embedding matrix by repeatedly concatenating v_a for N times through the operation $v_a \otimes I_N = [v_a, v_a, \ldots v_a]$, where I_N is a column vector with N dimensions. Then we append the aspect term embedding matrix into sentences embedding X as input of

LSTM$^{(a)}$ network, in this way, the output hidden representations H$^{(a)}$ can have richer aspect information:

$$H^{(a)} = LSTM^{(a)}([X, v_a \otimes I_N]) \tag{1}$$

Where $H^{(a)} = [h_{a1}, h_{a2}, \ldots, h_{aN}] \in \mathbb{R}^{d \times N}$, d is the number of hidden units, and N represent the length of the input sentence.

Similarly, the input of LSTM$^{(s)}$ network contains sentences embedding X and sentiment term embedding v_s. The sentiment term embedding matrix is obtained by repeatedly concatenating the sentiment term embedding v_s for N times through the operation $v_s \otimes I_N = [v_s, v_s, \ldots v_s]$. Then we append the sentiment term embedding matrix into sentences embedding X as input of LSTM$^{(s)}$ network, and the output hidden representations H$^{(s)}$ can have richer sentiment information:

$$H^{(s)} = LSTM^{(s)}([X, v_s \otimes I_N]) \tag{2}$$

Where $H^{(s)} = [h_{s1}, h_{s2}, \ldots, h_{sN}] \in \mathbb{R}^{d \times N}$.

3.2   Joint Attention Module

In order to capture the important aspect information and corresponding sentiment information of the sentence, we design a joint attention module by exploiting attention mechanism to concatenate the given aspect term embedding and sentiment term embedding with the feature representations of sentence respectively. The joint attention module contains aspect attention (as shown in (B-1) of Fig. 1) and sentiment attention (as shown in (B-2) of Fig. 1).

Aspect Attention. The aspect attention mechanism can model the aspect feature of sentence and obtain the inter-relation information between the given aspect terms and each word in the sentence. The input of aspect attention contains the hidden representation H$^{(a)}$ from LSTM$^{(a)}$ and the aspect term embedding v_a. We first extend the given aspect term embedding v_a to an aspect term embedding matrix through the operation $v_a \otimes I_N$, and take the aspect term embedding matrix as an aspect attention matrix. Then we calculate an aspect attention feature matrix A based on the concatenations of the aspect attention matrix and the output hidden representations H$^{(a)}$. Finally, the aspect attention weight vector r_a is obtained under the projection calculation of aspect attention feature matrix A:

$$A = tanh\left(W^{(a)}\left[H^{(a)}, v_a \otimes I_N\right]\right) \tag{3}$$

$$r_a = softmax\left(\left(u^{(a)}\right)^T A\right) \tag{4}$$

Where $A \in \mathbb{R}^{(d+d_w) \times N}$, $r_a \in \mathbb{R}^N$, $W^{(a)} \in \mathbb{R}^{(d+d_w) \times (d+d_w)}$ is a weight matrix, $u^{(a)} \in \mathbb{R}^{(d+d_w)}$ is a weight vector, $\left[H^{(a)}, v_a \otimes I_N \right]$ represent the concatenation of the output hidden representation and aspect term embedding matrix. The aspect weighted representation of sentence $h_{(a)}$ is obtained by weighting the hidden representation $H^{(a)}$ with aspect attention weight vector r_a:

$$h_{(a)} = H^{(a)} r_a^T \tag{5}$$

Sentiment Attention. The sentiment attention mechanism can capture the interrelation between the sentiment terms and aspect-related information in the sentence under the direction of sentiment terms information and contextual features of the words. The input of sentiment attention contains the hidden representation $H^{(s)}$ from the LSTM$^{(s)}$ and the sentiment embedding vector v_s. Similar as aspect attention, we extend the given sentiment term embedding v_s to a sentiment term embedding matrix through the operation $v_s \otimes I_N$, and take the sentiment term embedding matrix as a sentiment attention matrix to concatenate with the output hidden representations $H^{(s)}$ to calculate the sentiment attention feature matrix S. And the sentiment attention weight vector r_s is obtained under the projection calculation of sentiment attention feature matrix S:

$$S = tanh\left(W^{(s)} \left[H^{(s)}, v_s \otimes I_N \right] \right) \tag{6}$$

$$r_s = softmax\left(\left(u^{(s)} \right)^T S \right) \tag{7}$$

Where $S \in \mathbb{R}^{(d+d_w) \times N}$, $r_s \in \mathbb{R}^N$, $W^{(s)} \in \mathbb{R}^{(d+d_w) \times (d+d_w)}$ is a weight matrix, $u^{(s)} \in \mathbb{R}^{(d+d_w)}$ is a weight vector, $\left[H^{(s)}, v_s \otimes I_N \right]$ represent the concatenation of the output hidden representation and sentiment term embedding matrix. Then the sentiment weighted representation of sentence $h_{(s)}$ is obtained by weighting the hidden representation $H^{(s)}$ with sentiment attention weight vector r_s:

$$h_{(s)} = H^{(s)} r_s^T \tag{8}$$

Finally, the final sentence representation h^* is obtained by merging the aspect weighted representation of sentence $h_{(a)}$ with sentiment weighted representation of sentence $h_{(s)}$:

$$h^* = tanh\left(W_a h_{(a)} + W_s h_{(s)} \right) \tag{9}$$

Where $h^* \in \mathbb{R}^d$, $W_a \in \mathbb{R}^{d \times d}$ and $W_s \in \mathbb{R}^{d \times d}$ are the weight matrix parameters to be learned during the training.

3.3 Sentiment Classification Module

The joint attention module described in Sect. 3.2 have learned and took advantage of aspect term information and sentiment term information of sentence, making the final sentence representation h^* capture the important aspect feature and sentiment feature of the sentence. We use a softmax layer to output the probability of classifying the sentiment polarity of aspects:

$$y = softmax(W_b h^* + b) \tag{10}$$

Where $W_b \in \mathbb{R}^{d_c \times d}$ and $b \in \mathbb{R}^{d_c}$ are the parameters for softmax layer, d_c is the number of sentiment categories.

In order to train the proposed model better, we train and update the parameters of the model by backpropagation, and use the cross-entropy as loss function to measure the gap between the labeled sentiment distribution y and the predicted sentiment distribution $\hat{y}$. The model is optimized by minimizing the cross-entropy loss:

$$loss = -\sum_{i=1}^{D} \sum_{j=1}^{C} y_i^j log \hat{y}_i^j \tag{11}$$

Where D is the size of training dataset, C is the number of sentiment categories.

4 Experiment

We apply the proposed model to aspect-level sentiment classification with two different domain (Restaurant and Laptop) datasets to evaluate its effectiveness.

4.1 Experimental Setting

We implement the proposed model on the E5-2620CPUs workstation, the operating system is Linux (Ubuntu 14.04.4), the development environment is Python2.7 and TensorFlow1.4.0, and the development tool is PyCharm. As mentioned in Sect. 3.1, we train 300-dimension word embedding with the word2vec tool on two large domain-specific corpora (Yelp Challenge Dataset for restaurant domain, which contains 1.6 million reviews on restaurant, and Amazon Electronic Dataset for laptop domain, which contains 1.5 million reviews). For the unknown words in text, we initialize the word embedding with uniform distribution U (−0.01, 0.01). All the parameters except word embedding are initialized with normal distribution N (−0.01, 0.01). The size of hidden layer are 300 and AdaGrad optimizer is used to optimize the model with 0.01 learning rate.

4.2 Data Set

We evaluate our model on two public domain-specific datasets of SemEval2014 task 4, which include reviews from restaurant domain and laptop domain. The two datasets are labeled with fine-grained aspect-level human annotations and split into train and test set. Each review in the datasets is labeled with a list of aspect terms and the corresponding aspect terms polarities (positive, negative and neutral). Take the review "The price is reasonable although the service is poor" for an example, the aspect terms "price" and "service" and the corresponding aspect terms polarities "positive" and "negative" are labeled. Since the two datasets lack the sentiment terms annotations, in order to train our proposed model, we manually extract the adjective which corresponding to the aspect terms as sentiment terms label, that is, the "reasonable" and "poor" in the above example, and take the sentiment terms, aspect terms and the review text as input for model training phase and testing phase. For some review text which do not contain specific adjective corresponding to the aspect terms, we take the adjective "NULL" as the sentiment term annotations for the review text. The details of the two datasets are shown in Table 1, where "REST" represents the dataset of the Restaurant domain, and "LAPT" represents the dataset of the Laptop domain.

Table 1. The details of the two datasets, T represents the total number of sentences, A represents the total number of aspect terms, and S represents the total number of adjectives

Dataset		Positive	Negative	Neutral	Total		
					T	A	S
REST	Train	2064	798	537	3399	1219	1229
	Test	394	210	196	800	522	539
LAPT	Train	834	709	542	2085	955	1001
	Test	391	167	133	691	393	403

4.3   Comparison with Candidate Models

To demonstrate the effectiveness of our proposed joint attention LSTM network, we compare it with the following four state-of-the-art models that only attend to the given aspect terms information.

AT-LSTM: Attention-based LSTM (AT-LSTM) [16] uses LSTM network to learn the contextual feature of each word, and designs an attention mechanism to capture the aspect information. The aspect-specific sentiment information is extracted based on the aspect information and the contextual feature.

ATAE-LSTM: Attention-based LSTM with Aspect Embedding (ATAE-LSTM) [16] is an improvement of AT-LSTM, the model better make use of the aspect terms information by appending the aspect embedding into each word input vector. In such way, the LSTM network can take advantage of the aspect information for contextual feature extraction, leading the model to focus on the aspect information.

HEAT-BiGRU: Hierarchical attention network (HEAT- BiGRU) [17] uses BiGRU network to learn the contextual feature of each word, and the hierarchical attention module captures the aspect information of a sentence and uses the aspect information to capture the aspect-specific sentiment information.

HEAT-GRU: HEAT-GRU is a unidirectional version of HEAT-BiGRU, using GRU network to extract contextual features of each word and take the contextual features as input for the hierarchical attention module.

JAT-LSTM: JAT-LSTM is our proposed joint attention LSTM network model. The model make use of the aspect terms information and sentiment terms information, modeling the aspect feature and sentiment feature of a sentence, so that the final representations of the sentence can have the interdependence information between aspect terms and sentiment terms.

4.4   Experimental Results and Analysis

We evaluate the five different models' performance for sentiment polarity 2-class classification (positive, negative) and 3-class classification (positive, negative and neutral) based on the dataset of SemEval 2014 Task 4, the results are shown in Table 2.

Table 2. Accuracy on aspect-level sentiment classification for different models

Models	REST		LAPTOP	
	Bin.	Thr.	Bin.	Thr.
AT-LSTM	87.9	75.6	80.1	68.1
ATAE-LSTM	87.7	76.2	79.8	68.3
HEAT-GRU	87.9	84.3	80.0	84.2
HEAT-BiGRU	88.2	85.1	80.1	84.9
JAT-LSTM	**88.3**	**89.7**	**80.5**	**84.8**

Table 2 shows the performance of the five models on the Restaurant dataset and Laptop dataset, where "Bin." means 2-class classification and "Thr." means 3-class classification. For the performance of 2-class classification, our proposed JAT-LSTM network achieve better results than the other four models. For the performance of 3-class classification, the AE-LSTM and ATAE-LSTM that only attend to the given aspect terms information obtain lower classification accuracy on the two datasets. The HEAT-GRU and HEAT-BiGRU not only attend to the given aspect terms information, but also consider aspect-related information, and achieve higher classification accuracies. Our proposed JAT-LSTM network attend to both aspect terms information and sentiment terms information, and have achieved better performances.

It is notable that, on the LAPTOP dataset, our proposed model gets lower accuracy than HEAT-BiGRU. It is because the bidirectional model of HEAT-BiGRU can grasp both the former and the later contextual information of each word, and thus is powerful than the unidirectional model used on JAT-LSTM network. In addition, the improvements for laptop domain are not significant, and we conjecture the reason is that the joint attention LSTM model increases the number of parameter to learn, making it less applicable to small datasets. Since the proposed method shows good performance on two domain-specific datasets, the effectiveness of the joint attention LSTM network in aspect-level sentiment analysis tasks is verified.

5 Conclusions

In this paper, we propose a joint attention LSTM network for aspect-level sentiment analysis. This model take both the aspect terms information and sentiment terms information into consideration, and concatenate the aspect terms embedding and sentiment terms embedding with sentence embedding as the input of LSTM network to make the input information of the LSTM network richer. Then, the output hidden representation from the neural network is further computed with the aspect terms embedding and the sentiment terms embedding, leading the final sentence representation containing the aspect information and sentiment information. Finally, the final sentence representation is used for aspect-level sentiment classification. We evaluate our model with benchmark dataset from two different domains, and the experimental results show that our model obtain better performance than the baseline models which only consider the aspect terms information. Our model relies on manually labeled sentiment term, in the future, a more effective method of explicitly extract aspect terms and opinion terms will be considered to reduce the cost of heavy burden of manual annotation.

Acknowledgements. This work is supported by Chinese National Science Foundation (#61763007), Guangxi Natural Scicence Foundation (#2017JJD160017) and Innovation Project of GUET Graduate Education (#2018YJCX41).

References

1. Pang, B., Lee, L.: Opinion mining and sentiment analysis. Found. Trends® Inf. Retr. **2**(1–2), 1–135 (2008)
2. Liu, B.: Sentiment analysis and opinion mining. Synth. Lect. Hum. Lang. Technol. **5**(1), 1–167 (2012)
3. Yin, W., Schütze, H., Xiang, B., et al.: ABCNN: attention-based convolutional neural network for modeling sentence pairs. Computer Science (2015)
4. Lample, G., Ballesteros, M., Subramanian, S., et al.: Neural architectures for named entity recognition, pp. 260–270. arXiv:1603.01360 (2016)
5. Golub, D., He, X.: Character-level question answering with attention. arXiv:1604.00727 [cs. CL] (2016)
6. Li, Z., Zhang, Y., Wei, Y., et al.: End-to-end adversarial memory network for cross-domain sentiment classification. In: Proceedings of the International Joint Conference on Artificial Intelligence (2017)
7. Mnih, V., Heess, N., Graves, A.: Recurrent models of visual attention. In: Advances in Neural Information Processing Systems, pp. 2204–2212 (2014)
8. Bahdanau, D., Cho, K., Bengio, Y.: Neural machine translation by jointly learning to align and translate. Computer Science (2014)
9. Rush, A.M., Chopra, S., Weston, J.: A neural attention model for abstractive sentence summarization. Computer Science (2015)
10. Hu, M., Liu, B.: Mining and summarizing customer reviews. In: Proceedings of the Tenth ACM SIGKDD International Conference on Knowledge Discovery and Data Mining (ACM), pp. 168–177 (2004)

11. Yu, J., Zha, Z.J., Wang, M., et al.: Aspect ranking: identifying important product aspects from online consumer reviews. In: Proceedings of the 49th Annual Meeting of the Association for Computational Linguistics: Human Language Technologies, vol. 1, pp. 1496–1505. Association for Computational Linguistics (2011)
12. Nguyen, T.H., Shirai K.: Phrasernn: phrase recursive neural network for aspect-based sentiment analysis. In: Proceedings of the 2015 Conference on Empirical Methods in Natural Language Processing (2015)
13. Tang, D., Qin, B., Feng, X., et al.: Target-dependent sentiment classification with long short term memory. CoRR, abs/1512.01100 (2015)
14. Lakkaraju, H., Socher, R., Manning, C.: Aspect specific sentiment analysis using hierarchical deep learning. In: NIPS Workshop on Deep Learning and Representation Learning (2014)
15. Liang, B., Liu, Q., Xu, J., et al.: Aspect-based sentiment analysis based on multi-attention CNN. J. Comput. Res. Dev. **54**(8), 1724–1735 (2017)
16. Wang, Y., Huang, M., Zhao, L.: Attention-based lstm for aspect-level sentiment classification. In: Proceedings of the 2016 Conference on Empirical Methods in Natural Language Processing, pp. 606–615 (2016)
17. Cheng, J., Zhao, S., Zhang, J., et al.: Aspect-level sentiment classification with HEAT (HiErarchical ATtention) network. In: Proceedings of the 2017 ACM on Conference on Information and Knowledge Management (ACM), pp. 97–106 (2017)

Extraction New Sentiment Words in Weibo Based on Relative Branch Entropy

Liang Yang, Dongyu Zhang, Shaowu Zhang, Peng Fei,
and Hongfei Lin[✉]

School of Computer Science and Technology, Dalian University of Technology,
Dalian, China
{liang, zhangsw, hflin}@dlut.edu.cn

Abstract. There are a lot of new sentiment words appear in Weibo platform every day in the web2.0. As the unpredictable polarity of massive new words are detrimental to sentiment analysis for Weibo, hence how to extract new sentiment words and expand sentiment lexicon is of great importance. Therefore, we propose a relative branch entropy based method, which combines word frequency and adjacent words information to extract new sentiment word in Weibo. After integrated context and other factors, this method improves the accuracy of new sentiment word recognition. Three experiments are implemented on COAE 2014 Weibo corpus to compare the performance of different statistics with the proposed method. Experiment results show that the proposed method has a high accuracy, which demonstrates the effectiveness of this method and verify the promoted effect of new sentiment word extraction on the performance of Weibo sentiment classification.

Keywords: Sentiment word · Relative branch entropy · Sentiment analysis

1 Introduction

Weibo, namely Micro-Blog, is a social networking platform, on which users can share brief real-time information by setting up individual communities or following mechanism through WEB, WAP and other clients. Due to its short, real-time, high efficiency and other characteristics, it has become the dominant communication platform for the expression of ideas and views nowadays. By June 2015, the number of China's Weibo users was 204 million and the use rate of Net-surfers was 30.6% [1]. However, massive new words, especially new sentiment words, occur in Weibo frequently. These new sentiment words have brought great challenge to sentiment analysis, for example, the new word "狗带" as the homophony of "go die" means "go die", that if we don't know its negative emotion, which will have impacts to sentence-level sentiment analysis, such as some sentences similar to "今天的考试,我选择狗带(Today's exam, I choose to go die)".

As sentiment lexicon is the basis of sentiment analysis [2], some existing research firstly identify all the new words, then screen out new sentiment words, while there are also some other researchers extract new sentiment words directly. Among them, the statistics-based method are the most widely used because this approach does not rely

S. Zhang et al. (Eds.): CCIR 2018, LNCS 11168, pp. 158–169, 2018.
https://doi.org/10.1007/978-3-030-01012-6_13

on annotation. But After observed the statistical information, we found that large number of "new words" emerge with very flexible contexts frequently, such as "看事 (see thing)", "我来(I come)" and so on. Unfortunately, it is hard to find a suitable threshold to filter them, which affect the extraction accuracy of new sentiment words definitely. Hence, we propose a novel extraction method for new sentiment word in Weibo based on relative branch entropy. This method takes the context of words into account, which is able to filter out the incorrect neologisms with high frequency.

The main contributions of this paper are that we propose a novel statistics, namely relative branch entropy, to extract new sentiment words, and then set up detailed experiments to compare the performance with different statistics and the proposed method. After that, we further analyze different methods to determine the polarity of new sentiment word, experiment results provide an effective reference for related research.

The rest of the paper is structured as follows: we will introduce related word in the next section. We will emphatically describe the proposed method in Sect. 3. In Sect. 4, we will resent the experiments. Finally, the work is summarized in last Section.

2 Related Work

There are two main thoughts of extraction of new sentiment words at present. The first is to identify new sentiment words directly, for example, Hatzivassiloglou et al. [3] applied "and" or "but" and other conjunctions to obtain the polarity of connected adjectives through POS known words. Qiu et al. [4] employed syntactic analysis to conjecture and extend emotional word set. Huang et al. [5] started from very few seed words and used lexicon patterns to extend sentiment words and patterns iteratively in order to achieve final sentiment words. However, these methods not only requires construction of grammatical rules artificially, but also are ineffective to identify the noun in new sentiment words.

Another line is to identify all the new words firstly, and then screen out new sentiment words. For identifying the new words, there are a large number of researchers studying this issue from different point of view and methods, the methods are summarized as follows:

Rule-Based Methods. Such methods require manual summary and construct rules. Justeson et al. [6] used the regular expression to extract technical terminologies from documents. Isozaki et al. [7] proposed a simple-rule-based generator and decision tree learning method. Chen et al. [8] extracted Chinese new word employing morphological and statistical rules. Zheng et al. [9] established a rule base according to Chinese word formation for new word mining. However, these rule-based methods don't scale well, hard to maintain, and cannot exhaust all the linguistic phenomena.

Machine Learning Methods. Such methods threat new word identification as a machine learning binary classification problem, for example, Li et al. [10] proposed using an Independent Word Probability(IWP), term frequency in the document and so on as the features of SVM to classify candidate new words. Fu et al. [11] find new word boundary according to train the conditional probability model by using the POS context feature, joint model between words and word formation. Goh et al. [12] trained SVM model by using the Hidden Markov Model to annotate words to obtain character-based labels, and then detect new words by the character sequence. Li et al. [13] proposed applying neologisms pattern as the feature of SVM classifier, combined with rules to get new words. Xu et al. [14] also used SVM with related constraints and slack variables. However, machine learning models require not only heavy engineering of linguistic features, but also ex-pensive annotation of training data.

Statistics-Based Methods. Such methods focus on seeking the statistics of describing new word characteristics, like Pointwise Mutual Information [15], Probability Into Words [16], Rigidity [17] and so on. They select new word by setting the threshold. The first statistical model, Pointwise Mutual Information (PMI), was raised by Church et al. [15] in the 1990 in order to measure the degree of integration between words, since which statistics-based methods developed rapidly. Zhang et al. [18] proposed Enhanced Mutual Information (EMI), based on PMI, to measure the cohesion of co-occurrence word. Huang et al. [19] proposed Branch Entropy (BE) to measure the adjacent character uncertainty of candidate new words. Bu et al. [20] proposed Multi-word Expression Distance (MED) based on the information distance theory.

Then it is required a judgment of polarity for the identified new word. This problem can be considered as a classification problem which uses of existing seed lexicon or the polarity of document consist of candidate word, and the most of them utilized the term or synonym sets in Wordnet[1] [21, 22]. Other common approaches, based on the hypothesis that the similarity between the same polarity sentiment words, aim to calculate the similarity between candidate new words and known sentiment words, such as PMI based [23, 24], word embedding based [25], context similarity based [26] and so on. Among them, the word embedding based method used in this paper.

3   Methodology

In order to detect new words, it is need to use tools to segment Chinese word, so this paper uses Jieba[2] system, which deals with unknown words by HMM [27] and can find out some new words. However, it is inevitable to make some wrong segmentations for the massive emerging new words in the Internet.

To solve this problem, we preprocess Weibo corpus to obtain candidate word set, and use the relative branch entropy proposed to obtain the candidates. Then we train word embedding and compute the similarity between candidates and external sentiment lexicon [2], the larger the similarity is, the larger the probability of a candidate to be a sentiment word is. Finally, we discuss the results of new sentiment word extraction on the Weibo sentiment classification.

---

[1] http://wordnet.princeton.edu/.
[2] https://github.com/fxsjy/jieba.

3.1 Extraction of New Word Based on Relative Branch Entropy

The new word referred in this paper is the word not in the used lexicon, which contains 77,455 'old' sentiment words, which was provided by the Task 3 of COAE 2014. The new sentiment word is the new word, which contain emotions. We used the lexicon [2] to check the data, and obtained a number of 1,117,299 new words as candidate set *S*.

3.1.1 Rules

The set *S* contains some correct new words, such as "奥迪(Audi)", "给力(excellent)", but also contains a large number of incorrect ones, such as "宇大", "之末" and so on. This may be caused by some characters, namely stop-characters, which should appear alone rather than as a part of a word or phrase. Words contained stop characters are usually incorrect and can be filtered out. Therefore, we used Chinese stop-words list to extract out 264 stop-characters and constructed two types of stop-characters manually as shown in Table 1.

Table 1. Two types of stop-characters

Type	Amount	Examples
词首停用字 (Head-stop-characters)	165	已, 之, 此, 只, 虽
词尾停用字 (Tail-stop-characters)	171	哟, 吗, 了, 么, 呢

3.1.2 Improved Statistics Based on Branch Entropy

In order to find out the correct new words in *S*, we proposed an improved statistic method, named Relative Branch Entropy, which is based on Term Frequency and Branch Entropy.

Term Frequency (TF) is the number of the word appeared in corpus.

Branch Entropy (BE) is an important statistic, proposed by Huang et al. [19], for measuring the uncertainty of the adjacent characters of the candidate new word. The higher *BE* is, the higher uncertainty stands for. The *BE* of the new word is divided into **Left Branch Entropy (LBE)** and **Right Branch Entropy (RBE)**, and they are computed as follows:

$$LBE(w) = -\sum_l p(l|w) \log p(l|w)$$
$$RBE(w) = -\sum_r p(r|w) \log p(r|w) \tag{1}$$

Where *l* is the left adjacent character of *w*, *r* is the right adjacent character of *w*, $p(l|w)$ is the co-occurrence probability of *l* and *w*, then $p(r|w)$ is the co-occurrence probability of *r* and *w*.

Relative Branch Entropy

Only utilizing *TF* and *BE* is not enough to detect new word, for using *TF* can easily filter out some words whose *TF* equals 1 like "更会严(more strict)", but it cannot handle the high *TF* and incorrect words like "所行(doings)(*TF* = 22)", "看事(see thing) (*TF* = 16)" and so on.

Meanwhile, if we take *BE* into account, we can calculate out that *BE* of "所行 (doings)" is 2.1529, *BE* of "看事(see thing)" is 2.2200, so when the threshold of *BE* is set 2, the incorrect words could be filtered out. However, there are still a lot of new sentiment words whose *TF* are high and *BE* are low, such as "宅心仁厚(kind-hearted)"(*TF* = 10, *BE* = 0.3250), "淡然处之(take it lightly)" (*TF* = 21, *BE* = 0.1914). To solve the above mentioned problem, we proposed our method, named **Relative Branch Entropy** (*RBE*), and it computed as follows:

$$RBE(w) = \frac{TF(w)}{\min\{LBE(w), RBE(w)\}} \tag{2}$$

3.1.3 Algorithm

In a word, the algorithm of new word extraction is shown as follows:

Algorithm : **New word** extraction algorithm

Input :
　　S : **a set of** candidate word
　　HS : head-stop-characters
　　TS : tail-stop-characters
　　t_0 : the threshold of TF
　　t_1 : the threshold of BE
　　t_2 : the threshold of RBE

Output :
　　A list of new words, NewWords

1　for word in *S* :
2　　if the head-character of word in HS :
3　　　continue
4　　if the tail-character of word in TS :
5　　　continue
6　　if $TF(word) > t_0 \wedge BE(word) > t_1 \wedge RBE(word) > t_2$:
7　　　add word into NewWords
8　return NewWords

3.2   Extraction of New Sentiment Word Based on Word2Vec

Word2Vec[3] is a word embedding tool proposed by Mikolov et al. [28], which used of the context information to embed words to vectors. The similarity between words in semantics can be represented by calculating the distance of each word in the vector space. Based on this, we utilized the Skip-Gram model in Word2Vec to train the preprocessed data, and obtain vectors for all the words set V.

In order to identify new sentiment words in new words, we utilize the Affective Lexicon Ontology [2] as the seed word set, then the center sentiment vector $V_{sentiment}$ is computed as follows:

$$V_{sentiment} = \frac{1}{n} \sum_{i=1}^{n} v_i \tag{3}$$

Where v_i is the vector of seed word set.

After that, we can compute the similarity between new words and the center sentiment vector as follows to obtain the final new sentiment words.

$$Similarity(w|w \in NewWords) = \cos(V_w, V_{sentiment}) \tag{4}$$

Where V_w is the vector of w.

4   Experiment

4.1   Data Preparation and Evaluation Metric

In this paper, we will conduct three experiments, and detailed information of corpus is shown in Table 2. And we adapt P@N to evaluate the performance of extraction results for new sentiment word. P@N represents the accuracy of new sentiment word in top N words

Table 2. Information of corpus

Experiment	Corpus	Size	Number of Weibo	Average length	Data format
Experiment.1	COAE 2014 Task.3	1.55 GB	9,999,626	59.24 characters	<Doc_ID> 内容(content) (文字, #话题#, @用户名, 标点) </Doc_ID>
Experiment.2 Experiment.3	COAE 2014 Task.4	5.4 MB	40,000	54.18 characters	<Doc_ID> 内容(content) (文字, #话题#, @用户名, 标点) </Doc_ID>

---

[3] https://code.google.com/p/word2vec/.

4.2   Parameter Tuning

In order to obtain the thresholds of Algorithm, we firstly set $t_1 = 0$ and results are shown in Table 3, the performance of P@50 across different t_0 and t_2 settings.

Table 3. P@50 results for different t_0 and t_2

t_0	t_2							
	16	20	22	24	25	26	27	28
1	0.64	0.66	0.68	0.74	0.74	0.76	0.76	0.78
2	0.64	0.66	0.68	0.76	0.76	0.78	0.78	0.78
3	0.66	0.68	0.68	0.76	0.76	0.78	0.78	0.78
4	0.66	0.68	0.70	0.76	0.76	0.78	0.78	0.78
5	0.68	0.68	0.70	0.76	0.76	0.78	0.78	0.80

From the algorithm, we can know that the greater the threshold sets, the fewer candidate words get. Therefore, we choose the optimal setting as $t_0 = 2$, $t_2 = 26$.

Then we need to decide the optimal threshold of t_1. After setting $t_0 = 2$ and $t_2 = 26$, we annotate the top 100 words extracted, if the word is new sentiment word, the label is 1, else label is 0. The relevance of new sentiment word and **BE** shown in Fig. 2, where the vertical axis is the labels and the horizontal axis is the **BE** of each words.

From Fig. 1, we find there is no obvious relevance between whether a word is a new sentiment word and the number of *BE*, so it seems to be difficult to find an optimal threshold, and use it to separate new sentiment word between other words. However, when $t_1 = 0$, experiments achieve good results, hence t_1 set as 0 in this paper.

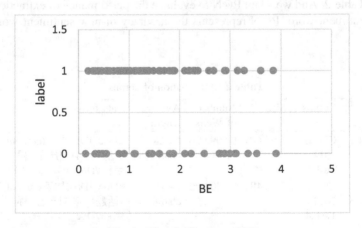

Fig. 1. The distribution of BE

4.3 Experiment Setting

In this section, we will conduct the following experiments:

Experiment. 1
We will compare our method with several baseline methods: PMI [15], EMI [18], MED [20], BE [19] and BA [29].

Experiment. 2
We will predict the polarity of new sentiment words. Firstly, three annotators labeled new sentiment words extracted by our proposed method, then the other two methods were adapted to predict the polarity of these words at the same time.

The first compared method is **Majority Vote** (**MV**), and the polarity is judged according to this rule: if $MV(w) > 0$, the word is positive; if $MV(w) < 0$, the word is negative; otherwise is neutral. **MV** is formulated as below:

$$MV(w) = \sum_{w_p \in P} \frac{count(w, w_p)}{|P|} - \sum_{w_n \in N} \frac{count(w, w_n)}{|N|} \tag{5}$$

Where P and N are the positive and negative set of the Affective Lexicon Ontology respectively, and $count(x, y)$ is co-occurrence times of input word x and y.

The second one is **Similarity of Vectors** (**SOV**), the judgement rule is the same as **MV**: if $P(w) > 0$, the word w is positive; if $P(w) < 0$, the word w is negative; otherwise is neutral. And **SOV** is computed as follows:

$$P(w) = \sum_{w_p \in P} \frac{\cos(V_w, V_{w_p})}{|P|} - \sum_{w_n \in N} \frac{\cos(V_w, V_{w_n})}{|N|} \tag{6}$$

Where V_x is the vector of word x.

Experiment. 3
In order to further justify whether expansion of new sentiment word would benefit sentiment classification on Weibo. We use the Weibo corpus in COAE 2014 (Task.4), the corpus consists of 40,000 Weibo posts, and there are 5,000 Weibo posts are sentimental. Firstly, we randomly sampled Weibo posts which contain at least one new sentiment word in the posts. Then two methods are applied for sentiment classification work.

The first method is a **lexicon-based model** (**Lexicon**), and its rule is that counts the number of positive and negative words respectively in each Weibo post, and then classifies the post to be positive or negative. Another method applies the **support vector machine** (**SVM**), where sentiment words are used as features, and 10-fold cross validation is conducted. The **Affective Lexicon Ontology** (**ALO**) [2] and ALO expanded with new sentiment words (denoted by **Expansion**) are utilized as sentiment lexicon resource respectively.

4.4   Results and Analysis

Experiment. 1

The results of different methods in experiment.1 are shown in Fig. 2. We can find that methods *BE* and *RBE* outperform the three baselines (PMI, EMI and MED) remarkably and consistently. From statistics perspectives, *BE* is better than other statistic-based methods. Meanwhile, for our *RBE* method integrate *TF*, *BE* and make full use of contextual statistic information of the candidate word, then the accuracy of *RBE* is higher than *BE*. For *BA*, which use the combination of adjacent strings as candidate words, while it ignores the correct new word before combination, which results in that *RBE* outperform *BA*.

As shown in Fig. 2, the P@N accuracy will decrease as *N* increase, mainly because the number of incorrect words (such as "不见泰山 (be shortsighted)", "欲加之罪(-condemn someone arbitrarily)"), are segmented from longer sentiment words as *N* increase. Due to these words are judged as non-sentiment words, which cause a worse accuracy. In addition, themes of the experiment corpus are complex, the polarities of some words are different in different domains, and this also affects the results.

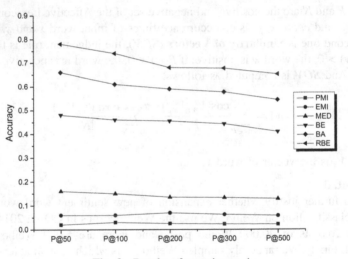

Fig. 2. Results of experiment. 1

Experiment. 2

In this experiment, we extracted top 200 new sentiment words from the similarity ranking list. Since there are three themes (jadeite, mobile phone and insurance) in the corpus, the annotators are requested to consider the domain dependent problem, for example, the word "起胶(light reflection continuously)" is not a sentiment word itself, but from jadeite perspectives, it is used to describe the jade is very valuable, so it should be labeled as positive. Similarly, the word "卡机(not running smoothly)" is also not a sentiment word itself, but it indicates a phone is not smooth, so it expresses a negative feeling in cell phone domain. Based on the above consideration, the label results shown as follows: (Table 4)

Table 4. Label results

Polarity	Amount	Example
Positive	89	起胶 (light reflection continuously), 佳品(treasure)
Negative	82	卡机 (not running smoothly), 骗保(Fraud)
Neutral	29	寒暄(exchange of conventional greetings), 缘分(fate)

Then, we apply *MV* and *SOV* for new sentiment word tow-classes and three-classes polarity classification respectively, the results is shown in Table 5. As we can see, the performance of *SOV* is much better than that of *MV*, because **SOV** is full use of contextual information by train vectors in neural network. However, the scale of corpus for training is small, so the improvement is not good enough. Another observation in three-classes polarity classification work is more difficult than two-classes polarity classification, for many extracted new sentiment words are nouns, and they are more hard to judge their polarities without domain knowledge.

Table 5. The results of two/three-class polarity classification

Number of classes	Methods	Accuracy
Two-classes	MV	0.845
	SOV	0.865
Three-classes	MV	0.520
	SOV	0.550

Table 6. The results of polarity classification of Weibo posts

Sentiment lexicon	Methods	Accuracy
ALO	Lexicon	0.657
	SVM	0.680
Expansion	Lexicon	0.705
	SVM	0.726

Experiment. 3

In this experiment, we randomly sampled 2,000 Weibo posts (Including 1,123 positive and 877 negative posts) that contain expanded sentiment word from the 5,000 Weibo posts that are official labeled. Results in Table 6 show that **SVM** model outperform **Lexicon** model generally, and expansion of new sentiment words improves the performance to a large degree, both models obtain 6–7% gains. It is an obvious proof for the effectiveness of extended sentiment words.

5 Conclusion

The sentiment lexicon forms the basis of sentiment analysis, due to the manually label work will take a lot of labor; hence it is impossible to cover the increasingly new words in Weibo all the time. Therefore, how to automatically extract new sentiment words from large-scale corpus and use it as an expansion of sentiment resources are great of importance to sentiment analysis researches.

In this paper, we propose a method to extract new sentiment word base on relative branch entropy (*RBE*). This method is full use of term frequency and adjacent information, and almost free of linguistic resources. Comparative experiments show that our proposed method outperforms than other baselines, which verify the effectiveness of the proposed method. Meanwhile, the experiments also demonstrate that expansion of sentiment words will benefit sentence-level sentiment classification.

However, there are still some problems unsolved, such as the wrong segmentations caused by longer words and the evaluation works. We will explore the solutions in our further researches.

Acknowledgements. This research is supported by the National Natural Science Foundation of China (No: 61702080, 61632011), the Fundamental Research Funds for the Central Universities (No. DUT17RC(3)016).

References

1. CNNIC: The 36th China Internet Network Development State Statistical Report. China Internet Network Information Center (2015)
2. Xu, L.H., Lin, H.F., Pan, Y., et al.: Constructing the affective lexicon ontology. J. China Soc. Sci. Tech. Inf. **27**(2), 180–185 (2008)
3. Hatzivassiloglou, V., McKeown, K.R.: Predicting the semantic orientation of adjectives. In: Proceedings of the Eighth Conference on European Chapter of the Association for Computational Linguistics. Association for Computational Linguistics, pp. 174–181 (1997)
4. Qiu, G., Liu, B., Bu, J., et al.: Opinion word expansion and target extraction through double propagation. Comput. Linguist. **37**(1), 9–27 (2011)
5. Huang, M., Ye, B., Wang, Y., et al.: New word detection for sentiment analysis. In: Proceedings of the 52th Annual Meeting on Association for Computational Linguistics. Association for Computational Linguistics, pp. 531–541 (2014)
6. Justeson, J.S., Katz, S.M.: Technical terminology: some linguistic properties and an algorithm for identification in text. Nat. Lang. Eng. **1**(01), 9–27 (1995)
7. Isozaki, H.: Japanese named entity recognition based on a simple rule generator and decision tree learning. In: Proceedings of the 39th Annual Meeting on Association for Computational Linguistics, pp. 314–321. Association for Computational Linguistics (2001)
8. Chen, K.J., Ma, W.Y.: Unknown word extraction for Chinese documents. In: Proceedings of the 19th International Conference on Computational Linguistics-Volume 1. Association for Computational Linguistics, pp. 1–7 (2002)
9. Zheng, J.H., Li, W.H.: A study on automatic identification for Internet new words according to word-building rule. J. Shanxi Univ. Nat. Sci. Edit **25**(2), 115–119 (2002)

10. Li, H., Huang, C.N., Gao, J., et al.: The use of SVM for Chinese new word identification. In: Conference First International Joint Conference on Natural Language Processing, pp. 723–732 (2004)
11. Fu, G., Luke, K.K.: Chinese unknown word identification using class-based LM. Lect. Notes Artif. Intell. **3248**, 704–713 (2005)
12. Goh, C.L., Asahara, M., Matsumoto, Y.: Machine learning-based methods to Chinese unknown word detection and POS tag guessing. J. Chin. Lang. Comput. **16**(4), 185–206 (2006)
13. Li, C., Xu, Y.: Based on support vector and word features new word discovery research. In: IEEE International Conference on Computer Science and Automation Engineering, pp. 287–294. IEEE (2012)
14. Yuanfang, X., Hui, G.: New word recognition based on support vector machines and constraints. In: 2nd International Conference on Information Science and Control Engineering (ICISCE), pp. 341–344. IEEE (2015)
15. Church, K.W., Hanks, P.: Word association norms, mutual information, and lexicography. Comput. Linguist. **16**(1), 22–29 (1990)
16. Chen, A.: Chinese word segmentation using minimal linguistic knowledge. In: Proceedings of the Second SIGHAN Workshop on Chinese Language Processing-Volume 17. Association for Computational Linguistics, pp. 148–151 (2003)
17. Wang, M.C., Huang, C.R., Chen, K.J.: The identification and classification of unknown words in Chinese: a N-grams-based approach. Logico-Linguist. Soc. Jpn. 113–123 (1995)
18. Zhang, W., Yoshida, T., Tang, X., et al.: Improving effectiveness of mutual information for substantial multiword expression extraction. Expert Syst. Appl. **36**(8), 10919–10930 (2009)
19. Huang, J.H., Powers, D.: Chinese word segmentation based on contextual entropy. In: Proceedings of the 17th Asian Pacific Conference on Language, Information and Computation, pp. 152–158 (2003)
20. Bu, F., Zhu, X., Li, M.: Measuring the non-compositionality of multiword expressions. In: Proceedings of the 23rd International Conference on Computational Linguistics, pp. 116–124. Association for Computational Linguistics (2010)
21. Kim, S.M., Hovy, E.: Determining the sentiment of opinions. In: Proceedings of the 20th International Conference on Computational Linguistics, p. 1367 (2004)
22. Esuli, A., Sebastiani, F.: Sentiwordnet: a publicly available lexical resource for opinion mining. In: Proceedings of LREC, vol. 6, pp. 417–422 (2006)
23. Volkova, S., Wilson, T., Yarowsky, D.: Exploring sentiment in social media: bootstrapping subjectivity clues from multilingual Twitter streams. In: Proceedings of the 51th Annual Meeting on Association for Computational Linguistics, pp. 505–510. Association for Computational Linguistics (2013)
24. Turney, P.D.: Thumbs up or thumbs down? Semantic orientation applied to unsupervised classification of reviews. In: Proceedings of the 40th Annual Meeting on Association for Computational Linguistics, pp. 417–424. Association for Computational Linguistics (2002)
25. Yang, Y., Liu, L.F., Wei, X.H., et al.: New methods for extracting emotional words based on distributed representations of words. J. Shandong Univ. Nat. Sci. **49**(11), 51–58 (2014)
26. Yu, H., Deng, Z.H., Li, S.: Identifying sentiment words using an optimization-based model without seed words. In: Proceedings of the 51th Annual Meeting on Association for Computational Linguistics, pp. 855–859. Association for Computational Linguistics (2013)
27. Huang, C., Zhao, H.: Chinese word segmentation: a decade review. J. Chin. Inf. Process. **21**(3), 8–20 (2007)
28. Mikolov, T., Chen, K., Corrado, G., et al.: Efficient estimation of word representations in vector space. arXiv preprint arXiv:1301.3781 (2013)
29. Chen, X., Wang, S.G., Liao, J.: Automatic identification of new sentiment word about microblog based on word association. J. Comput. Appl. **36**(2), 424–427 (2016)

Two-Target Stance Detection with Target-Related Zone Modeling

Huan Liu, Shoushan Li[(✉)], and Guodong Zhou

Natural Language Processing Lab, School of Computer Science and Technology,
Soochow University, Suzhou 215006, China
hliu0909@stu.suda.edu.cn,
{lishoushan, gdzhou}@suda.edu.cn

Abstract. Recently, stance detection on Twitter has been a hot research topic in the natural language processing community. Most previous studies have assumed that only one target is involved in a tweet and perform single-target stance detection. In this study, we address a more challenging version of this task, namely two-target stance detection, where two targets are involved. Specifically, we first define four categories of text zones related to two targets, i.e., *Target 1*, *Target 2*, *Non-target*, and *Other*, and propose an unsupervised approach to automatically obtain these zones. Then, we propose a hierarchical neural network to perform stance detection for each target where multiple LSTM layers are leveraged to encode the target-related zones and a Bidirectional LSTM layer to encode the outputs from the LSTM layers. Moreover, we introduce a target-related attention mechanism in the hierarchical network. Empirical studies demonstrate the effectiveness of the proposed approach to two-target stance detection.

Keywords: Stance detection · Twitter · Two-target · Attention mechanism

1 Introduction

Stance detection is the task of automatically detecting whether the author of a given text is in favor of, against, or neutral towards a target of interest. The target might be a product, person, organization, government policy and even an event. With the rapid development of many social platforms, such as Twitter and Facebook, studying stance detection could be beneficial for many real applications, such as identifying user attitudes towards a specific product and understanding public opinion towards a hot event. Over the last years, the natural language processing (NLP) community has become increasingly aware of the need for research on stance detection in many scenarios, such as congressional debates [1, 2], company-internal discussions [3], and debates in online forums [4, 5].

Recently, stance detection on Twitter has caused more and more interests. Specifically, SemEval-2016 organized a shared task on stance detection on Twitter and some studies have been carried their efforts on this task [6, 7]. Up to now, the proposed approaches to stance detection on Twitter have achieved some significant gains on the detection performances. However, almost all these studies are aimed at stance detection

© Springer Nature Switzerland AG 2018
S. Zhang et al. (Eds.): CCIR 2018, LNCS 11168, pp. 170–182, 2018.
https://doi.org/10.1007/978-3-030-01012-6_14

towards a single target. In contrast, stance detection towards two or more targets is also popular in real applications and existing approaches to single-target stance detection are not readily for this more challenging task, i.e., two-target stance detection, due to the following reasons.

Task: **Single-target Stance Detection**
E1: @JOEROWE409 @ScotsFyre Hillaey Clinton can't help but lie...it's who she really is. #SemST Target: *Hillary Clinton*   Stance: *AGAINST*
Task: **Two-target Stance Detection**
E2: *Stupid liberals who vote for #Hillary will cause Paris to happen in America. Only #Trump is telling the truth. #Trump2016 #HillaryClinton* Target1: *Donald Trump*   Stance1: *FAVOR* Target2: *Hillary Clinton*   Stance2: *AGAINST*

Fig. 1. An example of single-target stance detection and another example of two-target stance detection

First, most approaches to stance detection model it as a sentiment classification problem, and directly apply traditional text classification algorithms to perform stance detection, such as Mohammad [6] and Sobhani [8]. However, the sentiment of a tweet is sometime different from the stance in single-target stance detection, as illustrated in Mohammad [6]. In two-target stance detection, this problem becomes even more serious because there exists not only stance with opposite category to sentiment but also stance with mixed sentiment category, i.e., containing both *positive* and *negative* sentiment. For instance, as shown in Fig. 1, **E2** contains both a *negative* expression for *Hillary Clinton*, i.e., "*Stupid liberals who vote for #Hillary*" and a *positive* expression for *Donald Trump*, i.e., "*Only #Trump is telling the truth*".

Second, some related studies on single-target stance detection have noticed the above challenge and proposes some approaches to perform target-oriented sentiment classification [7]. However, in two-target stance detection, the length of the tweet becomes longer. Specifically, we calculate the sentence number in single-target and two-target stance detection corpora and find that the average sentence number is 1.4 in one tweet in the single-target corpus [6] but 2.1 in the two-target corpus [9] (It is named as multi-target corpus therein). In a longer tweet and two-target scenario, it might be much more difficult for the attention mechanism to capture the relationship between each word and each target.

In this paper, we focus on the two-target stance detection task and overcome the above difficulties by segmenting the tweet text into different target-related zones. Specifically, we first define four categories of target-related zones, i.e., *Target 1*, *Target 2*, *Non-target*, and *Other* and propose an unsupervised learning approach to detect these target-related zones. Then, we propose a hierarchical neural network where four LSTM models are leveraged to model the four zone texts and a Bidirectional LSTM is leveraged to further learn the outputs from the previous LSTM models. Moreover, to better model the target information, we introduce an attention mechanism in the hierarchical neural network. Empirical studies show that our approach outperforms several strong baselines.

The rest of our paper is structured as follows. Section 2 briefly discusses the related work. Section 3 gives the definition of the target-related zones and proposes our unsupervised learning approach to detect target-related zones in a tweet. Section 4 proposes our learning approach to two-target stance detection. Section 5 presents the empirical studies. Finally, Sect. 6 gives the conclusion and the future work.

2   Related Work

Stance detection has been widely studied in the NLP community and previous studies on this area have denoted their efforts to many types of text, such as online debating forums [5, 10, 11], congressional debates [1, 2], student essays [12], and microblog rumors [13]. More recently, with the development of SemEval-2016 shared task, stance detection on social media has drawn more and more attention, especially on the Twitter text [6].

Most previous studies on stance detection devote their efforts to finding more effective features and extra knowledge to improve the performances. Anand [4] derives both unigrams and bigrams as features and also use the Stanford parser to extract dependency features. Hasan [14] employs n-gram features and other complex features including dependency-based, frame-semantic, quotation and positional features. Besides searching the textual features, other studies aim to exploit extra information such as citation structure [15], rebuttal links [16], and sentiment information [17] to tackle this task. Among these studies, most of them conclude that simple features like word unigram or n-gram features are difficult to beat for stance detection [5, 6].

Some recent studies on stance detection devote their efforts on applying deep learning approaches to avoid spending too much substantial effort on hand-designing features. Augenstein [18] employs Bidirectional LSTMs with conditional encoding to weakly supervised stance detection and achieve comparable performance to the second best results reported in the SemEval 2016 shared task. Du [7] proposes a target-specific attention neural network to supervised stance detection and achieve the state-of-the-art performance.

However, all above studies solve stance detection towards single target in Twitter. Recently, Sobhani [9] originally builds a new Twitter dataset for multi-target stance detection and make the new dataset publicly available to facilitate further research. Furthermore, they propose several approaches to this novel task and conclude that the sequence-to-sequence neural network performs best among all approaches. Our work follows their work and focuses on two-target stance detection on Twitter.

3 Target-Related Zone Detection

3.1 Target-Related Zone Categories

A tweet consists of different parts of text for describing different targets. In this study, we refer to these textual parts related to different targets as target-related zones. Specifically, for a two-target scenario, we define four zone categories, namely *Target 1*, *Target 2*, *Non-Target* and *Other*.

(1). **Target 1:** This textual part talks about one target named *Target 1*.
(2). **Target 2:** This textual part talks about the other target named *Target 2*.
(3). **Non-Target:** This textual part talks about neither *Target 1* nor *Target 2*.
(4). **Other:** This textual part contains hashtags at the end of a tweet.

3.2 Our Approach to Target-Related Zone Detection

Target-related zone detection could be modeled as a two-step process where the first step is to segment the tweet into several small textual units, such as sentences and clauses, and the second step is to classify each textual unit into different target-related zone categories.

Formally, suppose that a text contains n words $w_i \in \mathbb{R}^d (1 \leq i \leq n)$, i.e., $x = \{w_1 w_2 w_3 \cdots w_n\}$, where d is the dimension of the word embedding. The objective of the first step, i.e., textual unit segmentation, is to group these words into several textual segments, i.e., $s = \{s_1 s_2 s_3 \cdots s_m\}(m \leq n)$, where $s_i = \{w_j w_{j+1} \cdots w_k\}$ $(1 \leq j \leq k \leq n)$. The objective of the second step, i.e., textual unit classification, is to classify each textual unit into each target-related zone category, i.e.,

$$f(s_i) = c$$
$$c \in \{Target1, \ Target2, \ Non\text{-}Target, \ Other\} \tag{1}$$

However, both textual unit segmentation and classification are nontrivial. Instead of annotating a large amount of labeled data for these two tasks, we employ an unsupervised approach to perform both textual unit segmentation and classification.

Specifically, for each target, the words in the target name are considered as target indicator substrings. For instance, for the target of *Hillary Clinton*, we use *hillary* and *clinton* as two target indicator substrings and regard a word as the target indicator if the word contains these substrings. Then, we leverage the target indicator to perform both textual unit segmentation and classification in three scenarios.

(1) **Tweet with no target indicator:** Given a tweet, if neither the *Target1* nor the *Target2* indicator could be found, the whole tweet is considered as one textual unit and this unit is classified to the category *Non-target*.
(2) **Tweet with only one target indicator:** Given a tweet, if only the *Target1* (or *Target2*) indicator could be found, the whole tweet is considered as one textual unit and this unit is classified to the category *Target 1* (or *Target 2*).

(3) **Tweet with two target indicators:** Given a tweet, if both the *Target1* and the *Target2* indicators could be found, following rules are employed to perform textual unit segmentation and classification.

Formally, the tweet is represented as follows

$$w_1 \cdots w_{T_1} \cdots w_{T_2} \cdots w_n \tag{2}$$

where w_{T_1} denotes the *Target1* indicator and w_{T_2} denotes the *Target2* indicator.

E3: *#hillaryclinton and #berniesanders wants people to keep a liberal victim mentality..for control ..#trump wants equality for all..#trump2016*
          Target1: *Donald Trump*   Target2: *Hillary Clinton*
Four Target-related Zones:
Target 1: *#trump wants equality for all..*
Target 2: *#hillaryclinton and #berniesanders wants people to keep a liberal victim mentality..for control ..*
Non-Target: *None*
Other: *#trump2016*

Fig. 2. An example and its target-related zones obtained by our detection approach

Then, we get three textual units, i.e.,

$$s_1 = \{w_1 \cdots w_{T_1}\} \tag{3}$$

$$s_2 = \{w_{T_1+1} \cdots w_{T_2-1}\} \tag{4}$$

$$s_3 = \{w_{T_2} \cdots w_n\} \tag{5}$$

The classification strategy is defined as follows:

$$f(s_1) = Target1 \tag{6}$$

$$f(s_3) = Target2 \tag{7}$$

$$f(s_2) = \begin{cases} Target1, & if\ l_1 \le l_3 \\ Target2, & if\ l_1 > l_3 \end{cases} \tag{8}$$

where l_1, l_2 and l_3 are the lengths of s_1, s_2 and s_3.

Figure 2 shows an example and its target-related zones obtained by our detection approach. From this figure, we can see that the target-related zones in this example are well detected.

4 Our Learning Model

4.1 Basic LSTM and Bidirectional LSTM Networks

Long Short-Term Memory Network (LSTM) is a special kind of Recurrent Neural Network (RNN) [19] and it aims to learn long-dependency correlations in a sequence. Specifically, LSTM has three gates: an input gate i_t, a forget gate f_t, an output gate o_t and a memory cell c_t. Each cell in LSTM can be calculated as follows:

$$i_t = \sigma(W_i x_t + U_i h_{t-1} + V_i c_{t-1}) \tag{9}$$

$$f_t = \sigma(W_f x_t + U_f h_{t-1} + V_f c_{t-1}) \tag{10}$$

$$o_t = \sigma(W_o x_t + U_o h_{t-1} + V_o c_{t-1}) \tag{11}$$

$$\widetilde{c}_t = \tanh(W_c x_t + U_c h_{t-1}) \tag{12}$$

$$c_t = f_t \odot c_{t-1} + i_t \odot \widetilde{c}_t \tag{13}$$

$$h_t = o_t \odot \tanh(c_t) \tag{14}$$

where x_t is an input vector at time step t, c_t denotes the LSTM memory, h_t is an output vector and the remaining are all sets of learned weight parameters. The symbol σ above is the sigmoid function and $\odot$ is the elementwise multiplication operation.

Bidirectional Long Short-Term Memory Network (Bi-LSTM) is also a RNN network which aims to efficiently make use of both past features (via forward states) and future features (via backward states) for a specific time frame [20]. Specifically, Bi-LSTM consists of two recurrent network layers, whereas the first one processes the sequence forwards $\overrightarrow{h_t}$ and the second one processes it backwards $\overleftarrow{h_t}$. The two hidden layer vectors are combined by formula (15):

$$h_t = \overrightarrow{h_t} \oplus \overleftarrow{h_t} \tag{15}$$

where $\oplus$ denotes the concatenation operation.

4.2 Hierarchical Networks

After target-related zone detection, the whole tweet is segmented into four zones, *Target 1*, *Target 2*, *Non-target* and *Other*. Figure 3 shows our neural network approach which accepts the four zones as inputs.

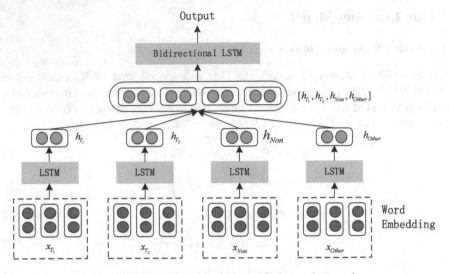

Fig. 3. The architecture of hierarchical neural network

Formally, the four zones are denoted as $x_{T_1}, x_{T_2}, x_{Non}, x_{Other}$. We use four LSTM layers to encode these four inputs, i.e.,

$$h_{T_1} = \text{LSTM}(x_{T_1}) \tag{16}$$

$$h_{T_2} = \text{LSTM}(x_{T_2}) \tag{17}$$

$$h_{Non} = \text{LSTM}(x_{Non}) \tag{18}$$

$$h_{Other} = \text{LSTM}(x_{Other}) \tag{19}$$

where h_{T_1}, h_{T_2}, h_{Non}, h_{Other} denotes the outputs of the four LSTM layers.

Then, we employ a merge layer to concatenate the four outputs from the LSTM layers and employ the Bi-LSTM model to learn the merging result, i.e.,

$$H = \text{Bi-LSTM}([h_{T_1}, h_{T_2}, h_{Non}, h_{Other}]) \tag{20}$$

where $H \in \mathbb{R}^{N \times (d+d)}$ is a matrix that the Bi-LSTM produced and N is the count of the target-related zone categories.

Finally, a softmax layer is employed to obtain the final classification result, i.e.,

$$y = \text{softmax}(W_s H + b_s) \tag{21}$$

where W_s and b_s are the parameters for the softmax layer. y is the conditional probability distribution over the three categories of stance, i.e., *Favor, Against*, and *None*, towards a target.

4.3   Attention-Based Hierarchical Networks

It is obvious that the target information is important for stance detection. Inspired by the recent work by Du [7], we introduce an attention mechanism to incorporate the target information. Note that their approach is applied in the scenario when the tweet is modeled as one input, which is not readily accessible for our task. Therefore, we propose our own attention-based approach, namely attention-based hierarchical network.

Figure 4 shows the architecture of attention-based hierarchical network. Compared to the hierarchical network above, there are two main differences. One is that the representation of each zone contains not only the word embedding but also the target embedding. The other difference is that an attention mechanism is applied after the Bi-LSTM layer. These two differences are introduced in the following subsection in detail.

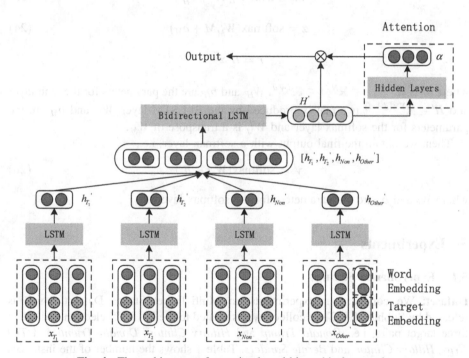

Fig. 4. The architecture of attention-based hierarchical networks

Target-Augmented Embedding

Similar to Du [7], we concatenate the embedding vectors of the text and target to learn a target-augmented embedding for modelling both text and target information. Formally, a target sequence containing N words is represented as $z = \{z_1 z_2 \cdots z_n\}$ and the target embedding is obtained by averaging the embedding of the words in the target sequence, i.e.,

$$\bar{z} = \frac{1}{N}\sum_{n=1}^{N} z_n \tag{22}$$

we augment the target embedding to the embedding of each word in the zones. Thus, the target-augmented embedding of a word w_i for a special target z is denoted as $w_i' = w_i \oplus \bar{z}$, $w_i' \in \mathbb{R}^{d+d}$, where $\oplus$ denotes the vector concatenation operation. Noted that the dimension of w_i' is $(d+d)$.

Therefore, the four zones are represented by both the words and target information, which are denoted as $x_{T_1}', x_{T_2}', x_{Non}', x_{Other}'$.

Attention-Based Network

The attention mechanism aims to learn an attention weight vector α and leverage the attention weights to form a new representation, i.e.,

$$M = \tanh(W_{H'}H' + b_{H'}) \tag{23}$$

$$\alpha = \mathrm{softmax}(W_M^T M + b_M) \tag{24}$$

$$r = H'\alpha^T \tag{25}$$

where $M \in \mathbb{R}^{N \times d}$, $\alpha \in \mathbb{R}^N$, $r \in \mathbb{R}^{d+d}$. $W_{H'}$ and $b_{H'}$ are the parameters for the tanh layer and $H' \in \mathbb{R}^{N \times (d+d)}$ is a matrix produced by the Bi-LSTM layer. W_M and b_M are the parameters for the softmax layer and W_M^T is a transpose of W_M.

Then, we obtain the final output with a softmax layer, i.e.,

$$y' = \mathrm{softmax}(W_r r + b_r) \tag{26}$$

where W_r and b_r are the parameters for the softmax layer.

5 Experiments

5.1 Experimental Settings

Dataset: We conduct our experiments on Multi-Target Stance Dataset which is released by Sobhani [9]. They collect tweets related to the 2016 US election and select three target pairs, i.e., *Donald Trump* and *Hillary Clinton*, *Donald Trump* and *Ted Cruz*, *Hillary Clinton* and *Bernie Sanders*. Table 1 shows the number of the instances in the training, development and test data sets in Multi-Target Stance Dataset.

Table 1. Distribution of the instances in Multi-Target Stance Dataset

Target pair	Train	Dev	Test	Total
Clinton-Sanders	957	137	272	1366
Clinton-Trump	1240	177	355	1722
Cruz-Trump	922	132	263	1317
Total	3199	446	890	4455

Word Embedding: We directly use the word embedding from Glove [21] where the vocabulary size is about 2.2 million and the dimension of each word embedding is 300.

Hyper-parameters: The dimension of LSTM layer output is 300. The optimization method is Adam and the learning rate is $5e - 4$, beta_1 is 0.9, beta_2 is 0.999 and epsilon is $1e - 8$.

Evaluation Metrics: The performance is evaluated using the average of the *F1-scores* for the two main categories, i.e., *Favor* and *Against,* which is defined as follows:

$$F_{Favor} = \frac{2P_{Favor}R_{Favor}}{P_{Favor} + R_{Favor}} \tag{27}$$

$$F_{Against} = \frac{2P_{Against}R_{Against}}{P_{Against} + R_{Against}} \tag{28}$$

$$F_{average} = \frac{F_{Favor} + F_{Against}}{2} \tag{29}$$

where P and R are precision and recall. Note that the final metrics does not consider the *None* category.

5.2 Results on Different Approaches to Two-Target Stance Detection

For thorough comparison, we implement the following approaches to two-target stance detection, i.e.,

- **LSTM:** The standard LSTM model including a LSTM layer, a fully connected layer and a dropout layer, without considering the information of target-related zones.
- **Target-specific attention neural network (TAN):** This is the state-of-the-art approach to single-target stance detection by Du [7]. In this approach, a target-specific attention neural network is trained to incorporate the target information.
- **Seq2Seq:** This approach employs the attention-based sequence-to-sequence model for two-target stance detection. This is the best-performed approach among several approaches, such as cascading SVMs and 9-class SVM, reported in the experiment by Sobhani [9]. Note that we directly use the performance reported in their paper and this result is comparable because our training, development, and test data are the same as theirs.
- **Hierarchical LSTM (H-LSTM):** This is our hierarchical network approach which leverages four target-related zones as inputs.
- **Attention-based Hierarchical LSTM (AH-LSTM):** This is our attention-based hierarchical network approach.

Table 2. Performances of our different approaches to two-target stance detection

Method	Target pair1		Target pair2		Target pair3		*Macro F-score*
	Trump	*Clinton*	*Trump*	*Cruz*	*Clinton*	*Sanders*	
LSTM	52.00	49.12	47.22	48.66	52.71	46.65	49.39
TAN	50.70	51.45	48.97	47.12	**52.87**	49.81	50.15
Seq2Seq	–	–	–	–	–	–	54.81
H-LSTM	**63.23**	54.94	48.36	**52.89**	46.54	56.94	53.82
AH-LSTM	61.98	**60.63**	**50.30**	52.88	52.26	**58.38**	**56.07**

Table 3. Performances of different approaches on the *None* category when the target is *Clinton* in Target pair1

Method	Category: *none*		
	P	R	F
LSTM	41.86	22.22	29.03
H-LSTM	75.47	49.38	59.70
AH-LSTM	**68.49**	**61.73**	**64.94**

Table 2 shows the performances of different approaches to two-target stance detection. From this table, we can see that TAN performs a bit better than LSTM, which highlights the importance of introducing the attention mechanism. Our approach H-LSTM performs much better than LSTM and TAN, which highlights the effectiveness of using the information of target-related zones. After using the attention mechanism, our approach AH-LSTM outperforms LSTM with a large margin, i.e., 6.68% in terms of *macro F-score*. Even compared to the strong baseline approach, Seq2Seq, our approach AH-LSTM is still superior. Note that, different from our approach, the Seq2Seq approach is a joint learning model, which jointly learns the stance detection tasks for both targets. This joint learning strategy could also be applied in our approach and we believe that this strategy would make our approach perform even better. This will be considered as our future work.

As mentioned above, the final metrics does not consider the *None* category. To better understand why our approach is so effective, we report the performances of different approaches on the *None* category. Due to the space limitation, we only report the results when the concerned target is *Clinton* in Target pair1, which are shown in Table 3. From this table, we can see that our approach performs extremely well in the *None* category, yielding an improvement of 35.91% over LSTM in *F-score*. This result once again highlights the importance of considering target-related zones in two-target stance detection.

5.3 Visualization of Attention

To better understand the attention mechanism in our approach, we visually analyze an example in the stance detection task for the target *Trump*. Figure 5 shows the attention visualization for four categories of target-related zones. Each line is a target-related zone produced by the approach in subsection 3.2. The color depth indicates the importance degree of attention weight for a target-related zone, the darker the more important. We can see that the attention mechanism can easily catch the important zone corresponding to the target *Trump*.

E4: *#BernieSanders #DonaldTrump Want to Stop The Looting of US Economy~#Hillary Just Whats To Get Even with Bill #msnbc #foxnews #nyt #Iowa #cnn*

Target 1:   #BernieSanders #DonaldTrump Want to Stop The Looting of US Economy~
Target2:   #Hillary Just Whats To Get Even with Bill
Non-Target: None
Other: #msnbc #foxnews #nyt #Iowa #cnn

Fig. 5. Attention visualization

6 Conclusion

In this paper, we propose a novel approach to two-target stance detection by incorporating the target-related zone information. Specifically, we first define four categories of target-related zones, i.e., *Target 1*, *Target 2*, *Non-target*, and *Other* and propose an unsupervised learning approach to detect these target-related zones. Then, we propose a hierarchical LSTM and its improved version with the attention mechanism, namely attention-based hierarchical LSTM, to perform stance detection for each target. Empirical studies show that our approach outperforms several strong baselines.

In our future work, we would like to manually annotate some labeled data on the target-related zones and propose a machine learning-based approach to improve the detection performance on the target-related zones. Moreover, we would like to combine the sequence-to-sequence model with our attention-based hierarchical LSTM to further improve the performance on two-target stance detection.

Acknowledgments. This research work has been partially supported by two NSFC grants, No. 61672366 and No. 61331011.

References

1. Thomas, M., Pang, B., Lee, L.: Get out the vote: determining support or opposition from congressional floor-debate transcripts. In: Proceedings of EMNLP, pp. 327–335 (2006)
2. Yessenalina, A., Yue, Y., Cardie, C.: Multi-level structured models for document-level sentiment classification. In: Proceedings of EMNLP, pp. 1046–1056 (2010)
3. Agrawal, R., Rajagopalan, S., Srikant, R., Xu, Y.: Mining newsgroups using networks arising from social behavior. In: Proceedings of WWW, pp. 529–535 (2003)

4. Anand, P., Walker, M., Abbott, R., Tree, J.E.F., Bowmani, R., Minor, M.: Cats rule and dogs drool!: classifying stance in online debate. In: The Workshop on Computational Approaches to Subjectivity and Sentiment Analysis, pp. 1–9 (2011)

5. Somasundaran, S., Wiebe, J.: Recognizing stance in ideological online debates. In: The Workshop on Computational Approaches to Analysis and Generation of Emotion in Text, pp. 116–124 (2010)

6. Mohammad, S.M., Kiritchenko, S., Sobhani, P., Zhu, X., Cherry, C.: Semeval-2016 task 6: detecting stance in tweets. In: The Workshop on Semantic Evaluation (SemEval-2016), pp. 31–41 (2016)

7. Du, J., Xu, R., He, Y., Gui, L.: Stance classification with target-specific neural attention networks. In: Proceedings of IJCAI, pp. 3988–3994 (2017)

8. Sobhani, P., Mohammad, S.M., Kiritchenko, S.: Detecting stance in tweets and analyzing its interaction with sentiment. In: Joint Conference on Lexical and Computational Semantics, pp. 159–169 (2016)

9. Sobhani, P., Inkpen, D., Zhu, X.: A dataset for multi-target stance detection. In: Proceedings of EACL, pp. 551–557 (2017)

10. Somasundaran, S., Wiebe, J.: Recognizing stance in online debates. In: Proceedings of ACL and AFNLP, pp. 116–124 (2009)

11. Hasan, K.S., Ng, V.: Stance classification of ideological debates: data, models, features, and constraints. In: Joint Conference on Natural Language Processing, pp. 1348–1356 (2013)

12. Faulkner, A.: Automated classification of stance in student essays: an approach using stance target information and the Wikipedia link-based measure. Science 376(12), 86 (2014)

13. Qazvinian, V., Rosengren, E., Radev, D.R., Mei, Q.: Rumor has it: identifying misinformation in microblogs. In: Proceedings of EMNLP, pp. 1589–1599 (2011)

14. Hasan, K.S., Ng, V.: Why are you taking this stance? Identifying and classifying reasons in ideological debates. In: Proceedings of EMNLP, pp. 751–762 (2014)

15. Burfoot, C., Bird, S., Baldwin, T.: Collective classification of congressional floor-debate transcripts. In: Proceedings of ACL, pp. 1506–1515 (2011)

16. Walker, M.A., Anand, P., Abbott, R., Grant, R.: Stance classification using dialogic properties of persuasion. In: Proceedings of ACL, pp. 592–596 (2012)

17. Ebrahimi, J., Dou, D., Lowd, D.: A joint sentiment-target-stance model for stance classification in tweets. In: Proceedings of CL, pp. 2656–2665 (2016)

18. Augensten, I., Rocktaschel, T., Vlachos, A., Bontcheva, K.: Stance detection with bidirectional conditional encoding. In: Proceedings of EMNLP, pp. 876–885 (2016)

19. Hochreiter, S., Schmidhuber, J.: Long short-term memory. Neural Comput. 9(8), 1735–1780 (1997)

20. Schuster, M., Paliwal, K.K.: Bidirectional recurrent neural networks. IEEE Trans. Sig. Process. 45(11), 2673–2681 (1997)

21. Pennington, J., Socher, R., Manning, C.D.: Glove: global vectors for word representation. In: Proceedings of EMNLP, vol. 12, pp. 1532–1543 (2014)

Social Computing

Information Diffusion Model Based on Opportunity, Trust and Motivation

Jihong Wan[1] , Xiaoliang Chen[1,2](✉) , Yajun Du[1] , and Mengmeng Jia[1]

[1] School of Computer and Software Engineering,
Xihua University, Chengdu 610039, China
chexiaol@iro.umontreal.ca

[2] Department of Computer Science and Operations Research,
University of Montreal, Montreal, QC H3C3J7, Canada

Abstract. Building an accurate information diffusion model around universal social factors has started to post its popularity on social network researches, which benefits a lot from its evolution simulation for identifying the messages with better prices, promoting news online quickly, and controlling public opinions. This paper constructs a novel model with respect to combine three factors: Opportunity, Social Trust and Game Selection Motivation. Firstly, the interest similarity between two users is convenient for measuring the opportunity to receive a message. Secondly, the threshold of social trust is calculated by coupling users network influence and content contribution. Thirdly, game selection with a rule to compute the best benefits has recognized as the motivation of users to spread a message. Finally, this paper presents an improved page rank algorithm to build a model by the idea of the game selection based on opportunity and social trust. Experimental result shows that social trust can accelerate the spread of information in Microblog social network by considering both the network topology and information content simultaneously.

Keywords: Social network · Information diffusion · Social trust
Game selection

1 Introduction

Online communities such as Facebook, Twitter and Microblog have become increasingly popular among people for information communication and social

Supported by the National Natural Science Foundation (Grant No. 61472329, 6160 2398), the Innovation Fund of Postgraduate, Xihua University (No. ycjj2017176), the Students' Platform for Innovation and Entrepreneurship Training Program (No. 2018069), the Chunhui Plan Cooperation and Research Project, Ministry of Education of China (Z2015100, No. Z2015109), Scientific Research Fund of Sichuan Provincial Education Department (No. 15ZA0130), the Key Scientific Research Fund of Xihua University (No. z1412616, No. z1422615) and the Open Research Subject of Key Laboratory of Security Insurance of Cyberspace, Sichuan Province (No. szjj2015-057).

© Springer Nature Switzerland AG 2018
S. Zhang et al. (Eds.): CCIR 2018, LNCS 11168, pp. 185–197, 2018.
https://doi.org/10.1007/978-3-030-01012-6_15

contacts. It is widely acknowledged that everybody has been making possible effort to post their opinions in their focused virtual communities. However, the very invention has also created some potential problems that almost everybody is exposed to. For example, an increasing number of people have become obsessed with transferring unverified gossip without doing anything meaningful in virtual communities. Hence, developing a highly accurate information diffusion model benefits a lot for studying potential information diffusion processes in social media. Perfect examples can be found in such well-recognized information flow analysis as the prediction and the guide of public opinion in social networks.

The topic of information diffusion model in social networks has been widely focused. However, there has been an excessive abstraction in existing work during study the information diffusion problem, ignoring some significant social, motivational and affective factors [1]. Consequently, this paper presents three kinds of social factors that affect the opportunity, trust and motivation of information diffusion in social networks. Then, evolutionary game theory [2,3] is introduced in this paper to construct a model with hybrid social factors.

A new model with hybrid social factors is established according to the network topology, the content of microblog and the information of users. The content of this paper is organised as follows. Firstly, the relative work of microblog social network and information diffusion models is presented. Secondly, three types of social factors are described as the opportunity, social trust and game selection motivation, which measures are defined as:

1. **Opportunity of receiving information:** the interest similarity between adjacent users, which is calculated by an improved Tanimoto coefficient [4].
2. **The threshold value of social trust:** the coupling value of user's network influence and content contribution.
3. **User's choice motivation for information diffusion:** the characteristics of maximized individual interests in evolutionary games.

Finally, Game selection based on Opportunity and Trust-Page Rank (OT-Page Rank) algorithm is proposed, which is used to construct the information diffusion model of hybrid social factors based on opportunity, trust and motivation. The validity of the model is illustrated by experiments.

2 Related Work

2.1 Microblog Social Network

As an important platform for users to obtain information and social contacts in a timely manner, microblog social network is usually used as an effective and realistic medium for analyzing and verifying the laws of information diffusion [5,6]. Through the introduction of the relationship of human interaction in the real world, the social relationship among users has become an important factor affecting the effectiveness of information diffusion. In addition, some social factors that are specific to social people, such as motivation and trust, can be reflected in microblog social network. Therefore, we can effectively control and guide the network public opinion that threatens social security and causes social unrest by studying the process of information diffusion in microblog social network.

2.2 Information Diffusion Models

Social network analysis relies on the construction of information diffusion model according to the actual communication mechanism, in which the communication mechanism and the sharing relationship mechanism in the social network complement each other. The research on information diffusion models related to this article is mainly divided into the following three aspects:

(1) *Information diffusion models based on diffusion process*

The researchers proposed that the information diffusion of online social network has the same pattern as the epidemic spread of the real world. Most of the information diffusion models originate from the classical SI, SIS, SIR and SIRS epidemic models. Daley and Kendall [7] first proposed the Daley-Kendall model of rumor spreading by analyzing the relationship between infectious disease diffusion and information diffusion. Later, the Daley-Kendall model was widely used in various information diffusion models [8]. These information diffusion models mainly focus on the dynamic of the diffusion process and the redistribution of individuals. But they do not take into account some key social factors that users themselves have in the process of information diffusion, and do not show the key characteristics of the diffusion.

(2) *Information diffusion models based on network key nodes*

The "opinion leader" as a key user was originally proposed by Lazarsfeld et al. [9] in 1950s. Various opinion leaders emerged in response to emergencies can influence and guide the decision-making choices of the public, and thus play an important role in information diffusion [10]. Yong et al. [11] evaluated the influence of opinion leaders through trust indicators and distrust indicators between users. Ortega et al. [12] proposed a user polarization trust mechanism based on trust and reputation system to achieve diffusion in social network through positive and negative link scores. Although these researchers use the trust mechanism to determine the key role of opinion leaders in the process of information diffusion, they do not pay attention to the relationship and importance of the trust mechanism between opinion leaders and users' information diffusion.

(3) *Information diffusion models based on evolutionary game*

Some scholars analyzed the information forwarding behavior from the perspective of game theory. Zinoviev et al. [2]. used the knowledge level, the degree of trust and influence of the actor to form the publishing and comment strategies of the network users. And a game theory model of the one-way information forwarding and feedback mechanism of the star-type social network was proposed.

Qiu et al. [3]. used the utility function to combine the user's own features (such as belief, persuasion and reputation, etc.) with the user's behavior characteristics (diffusion and reception) to obtain the benefit of the user's game in the process of information diffusion. These studies only unilaterally regard trust as a measure of user behavior. However, they do not take into account the key bridge role that it play in the whole process of information diffusion in the network topology.

3 Opportunity, Trust, and Game Motivation

This section defines the microblog social network, opportunity, trust, and game motivation, respectively.

A microblog social network is a directed weighted graph $SN = (V, E, \Delta, \Phi)$, where $V = \{\mu_1, \mu_2, \ldots, \mu_n\}$ and $E = V \times V = \{\varepsilon_{ij}\}$ are the set of users with size n and the similar association of interests between users, respectively; $\Delta = \{\delta_{u_i}^\omega\}$ $i = 1, 2, \ldots, n$ is the weight set of users; $\Phi = \{\varphi_{\varepsilon_{ij}}^\omega\}$ $i, j = 1, 2, \ldots, n$ is the weight set of associated edge between the microblog users, which measures the chance of microblog users to receive a certain information.

(1) *Opportunity*

Neighbor users with the similar interest of a user u_i may focus on a same topic. Neighbor users hence have an opportunity to receive the message that u_i posts on microblog. According to the interest characteristics of microblog users, focused interests for a user are usually presented as a Boolean vector. Then, improved Tanimoto similarity coefficient can be used to calculate the similarity between individuals in Boolean metric.

Definition 1 (Opportunity). The opportunity of user u_j to receive the message m transmitted by user u_i is defined by the following formula:

$$Opp_{\langle u_i, u_j, m\rangle, u_j \in N(u_i)} = \frac{\overrightarrow{u_i}' \bullet \overrightarrow{u_j}'}{\left|\overrightarrow{u_i}'\right|^2 + \left|\overrightarrow{u_j}'\right|^2 - \overrightarrow{u_i}' \bullet \overrightarrow{u_j}'} = \varphi_{\varepsilon_{ij}}^\omega \qquad (1)$$

where, m represents a certain message that is spreading on the microblog and $N(u_i)$ is the adjacent nodes of the user u_i. Notation $\overrightarrow{u_i}$ represents the interest vector of the user u_i and $\overrightarrow{u}_i' = \{\overrightarrow{u}_{i1}', \overrightarrow{u}_{i2}', \ldots, \overrightarrow{u}_{ik}'\}$ with size $k = Count(\overrightarrow{u}_i \cup \overrightarrow{u}_j)$ is its Boolean dimension representation. Operator $Count(vector\ \varepsilon)$ calculates the number of elements in vector ε. Let χ_k represents a interest in interest vector $\overrightarrow{u}_i$. Then, an element $\overrightarrow{u}_{ik}'$ of $\overrightarrow{u}_i'$ equals 1 if $\chi_k \in \overrightarrow{u}_i'$, in which $\chi_k \in \overrightarrow{u}_i \cup \overrightarrow{u}_j$, otherwise, $\overrightarrow{u}_{ik}'$ equals 0. This formula finally obtains a opportunity weight value of the associated edge between any pair of microblog users $\varphi_{\varepsilon_{ij}}^\omega$.

(2) *Social trust*

Determining the credibility of a message is necessary for a microblog user when he/she is likely to receive a message from their neighbors, which is usually depended on personal knowledge and thereby creating user feedback behaviors such as reading, sharing, commenting, and praising. Developing a reasonable measure of social trust contributes a lot to predict users' actual feedback behavior. This paper measures the credibility of users through two dimensions: network topology and content contribution.

A. *Network topology dimension*

The influence of information content cannot be measured according to network topology. Hence, the network influence [13] of a user who posts a message is used in this work as one of standards of user's social trust.

Definition 2 (Network Opinion Leader,NOL). The network influence of user u_i is defined as:

$$NOL(u_i) = \frac{deg^{in}(u_i)}{\sum_{i=1}\sum_{j>1}\varepsilon_{ji}} \tag{2}$$

Definition 2 shows the ratio of node u_i's in-degree to the number of edges between the nodes in the layer with u_i and the nodes belonging to the upper layer.

B. *Content contribution dimension*

This work measures the content contribution of users by the frequency of posting microblogs, the amount of microblog posts, and the number of users' feedbacks. User behavior relationships are listed as follows to explain the feedback.

- Microblog behavior 'share', denoted by $ms(u_{j\langle i,j\rangle\in\varepsilon_{ij}}) = \langle m_i, m_j, s\rangle$, is defined as the relationship between two microblog messages, where m_i represents the posted original microblog messages by user u_i, and $\langle m_i, m_j, s\rangle$ represents the sharing relationship between two users. The information diffusion direction is $u_i \dashrightarrow u_j$ if user u_j shares user u_i's microblog information.
- Microblog behaviors 'comment' and 'praise' are represented by the relationship between users and messages. Different contributions of 'comment' and 'praise' are expressed as: a 'comment' $mc(u_{j\langle i,j\rangle\in\varepsilon_{ij}}) = \langle u_j, m_i, c\rangle$ of the user u_j to the user u_i's message m and a 'praise' $mp(u_{j\langle i,j\rangle\in\varepsilon_{ij}}) = \langle u_j, m_i, p\rangle$ of the user u_j to the user u_i's message m.

Definition 3 (Feedback). In microblog social network, the feedback of user u_i obtained from other users $MFB(u_i)$ is organized as follows:

$$MFB(u_i) = \alpha_1 MS(u_i) + \alpha_2 MC(u_i) + \alpha_3 MP(u_i) \tag{3}$$

Among them, operations $MS(u_i) = \sum_j ms(u_{j\langle i,j\rangle\in\varepsilon_{ij}})$, $MC(u_i) = \sum_j mc(u_{j\langle i,j\rangle\in\varepsilon_{ij}})$, and $MP(u_i) = \sum_j mp(u_{j\langle i,j\rangle\in\varepsilon_{ij}})$ represent the total number of 'sharings', 'comments', and 'praisings', respectively.

Definition 4 (Content Contribution). The degree of content contribution of user u_i in the process of information diffusion $CC'(u_i)$ is expressed as:

$$CC'(u_i) = \frac{CC(u_i) - min[CC(u_i)]}{max[CC(u_i)] - min[CC(u_i)]} \tag{4}$$

$$CC(u_i) = \alpha_4 MF(u_i) + \alpha_5 MN(u_i) + MFB(u_i) \tag{5}$$

Among them, the content contribution of the microblog user u_i $CC(u_i)$ is determined by three parts: the frequency of publishing microblog posts within a specified period of time $MF(u_i)$, the amount of microblog posts $MN(u_i)$, and the number of user's feedbacks $MFB(u_i)$ according to certain message. Finally, normalized the content contribution.

According to Definitions 2 and 4, user's network influence and content contribution are combined to measure the social trust of user in microblog.

Definition 5 (Social Trust). The social trust $ST(u_i)$ of user u_i is defined as:

$$ST(u_i) = \eta NOL(u_i) \times (1 - \eta)CC'(u_i) = \delta_{u_i}^{\omega} \tag{6}$$

Microblog users generate the motivations of sharing, commenting, and praising based on social trust, then carry out the next step of information diffusion. On the other hand, users always choose the messages with their best interest to share in the whole net. The motivation of choose can be described by selection strategies in game theory.

(3) *Game selection motivation*

The motivation of participants has been playing an important role in ensuring the reliable information diffusion process in social networks. This section uses the Nash equilibrium theorem, the core point in game theory, to describe the motivation. An example can be found in the user decisions on features such as 'sharing' and 'praising' in microblog. The final decisions of users are quite sensitive to their own benefits influenced by their neighbors. Obviously keeping maximal benefits by game selection will contribute a lot to run a well realistic diffusion model.

Table 1. Microblog user information diffusion revenue matrix.

	$T_1(Share)$	$T_0(unShare)$
$T_1(Share)$	$b_{u_1 u_2}^{T_1 T_1}$	$b_{u_1 u_2}^{T_1 T_0}$
$T_0(unShare)$	$b_{u_1 u_2}^{T_0 T_1}$	$b_{u_1 u_2}^{T_0 T_0}$

We consider the game between a user and its multiple neighbors as the mathematics game model combined with many two-person games. No any other individual can increase its income by unilaterally changing best strategy. Therefore, the Nash equilibrium theorem can be utilized to obtain the necessary decision distribution of users in the stable state of information diffusion. Table 1 shows the revenue matrix on information diffusion in microblog. Notation T_1 and T_0 represent the operation 'sharing' and not 'sharing', respectively. The measure value of user's income can be calculated by the coupling method of computing opportunity and social trust.

On the basis of opportunity and social trust, microblog users have such behavior motivations as 'sharing', 'comment', 'praising', and 'retransmission' by their own revenue games so as to spread the information. Therefore, the game motivation is defined as follows:

Definition 6 (Game Motivation). The behavior motivation between two users u_1 and u_2 at a certain moment is defined as the greatest benefit of the game among behaviors:

$$Max\{f|f : (b_{u_i u_j}^{\Theta(\tau)}) \to (b_{u_i u_j}^{\Theta(\tau)})'\} \tag{7}$$

Among them, the value 0 (resp. 1) in the set $\tau = \{0,1\}$ means that no messages are shared (resp. share messages). No any benefit appears if the value equals 0. Set $\Theta(\tau) = \{T_x T_y | \forall x, y \in \tau\}$ represents the choice of users to share information or not to share information. The benefit accepted by microblog users can be considered as the mapping $f : (b_{u_i u_j}^{\Theta(\tau)}) \to (b_{u_i u_j}^{\Theta(\tau)})'$ from the initial benefit distribution $(b_{u_i u_j}^{\Theta(\tau)}) = \delta_{u_i}^{\omega} \times \varphi_{\varepsilon_{ij}}^{\omega}$ to the steady state $(b_{u_i u_j}^{\Theta(\tau)})'$. Both distributions are set and calculated by Game selection based on Opportunity and Trust Page Rank algorithm. Consequently, users will choose the most beneficial results by game selection as the next receiver for information diffusion.

4   Information Diffusion Model Based on Opportunity, Trust and Motivation

This section uses the opportunity and social trust to calculate the initial-state probability distribution value, and adopts Page Rank continuously iterated to steady-state probability distribution value that is the information income of each nodes. Then, the user node with the largest benefit can be obtained as the result of the game selection for upper level users, which is the next receiver. Hence, a diffusion model can be constructed by iterating the selection process.

Page Rank [14] may serve as an effective way to analyze Web links without considering the content of queries. A score within $[0, 1]$ is added to each node according to the Web graph. OT-Page Rank updates the initial probability distribution by coupling both vector values of opportunity and social trust at each stages of information diffusion. In each stage, a focused user that selects which user to send a message is determined by the opportunity of received message from the users in the next layer and the trust to the message. Opportunity is here denoted by the edge weight between two adjacent nodes. Social trust is denoted by the weight of user node with initial value 1.

This paper assumes that the interest similarity between two users is bidirectional and hence $\gamma = 0.5$. OT-Page Rank value of a user u_i can be calculated as follows:

$$OT - PR(u_i) = \frac{1-\gamma}{N} + \gamma \sum_{u_j \in M_{u_i}} \frac{UR(u_j)}{deg^{(out)}(u_j)} \tag{8}$$

Among them, M_{u_i} is a user set that have outgoing links to user u_i, $deg^{(out)}(u_j)$ denotes the number of outgoing links of user u_j, and N is the number of users.

Algorithm 1. Game selection based on OT-Page Rank.

Require:
  Microblog social network $SN = (V, E, \Delta, \Phi)$;
  $\gamma = 0.5$; $\alpha_1, \alpha_2, \alpha_3, \alpha_4, \alpha_5, \eta_1, \eta_2$.
Ensure:
  Information diffusion process map SN'.
1: Initial map $SN' = (V', E', \Delta', \Phi')$;
2: **for** each $i \in length[uid\_list]$ **do**
3:    According to the formula (1), calculate the opportunities for users to receive information
      $Opp_{\langle u_i, u_j, m \rangle}, u_j \in N(u_i)$ (denoted by $\varphi^\omega_{\varepsilon_{ij}}$);
4:    Calculate the network influence of users NOL based on formula (2);
5:    Calculate information contribution $CC(u_i)$ for each user by using the microblog user infor-
      mation crawled in conjunction with equations (3) and (5);
6:    Normalize the content contribution of user by the formula (4);
7:    Calculate the social trust of microblog users $ST(u_i)$(denoted by,$\delta^\omega_{u_i}$) based on formula (6);
8:    Compute initial probability distribution vector by OT-Page Rank algorithm $\vec{x}_0 = (b^{\Theta(\tau)}_{u_i u_j}) \leftarrow$
      $\varphi^\omega_{\varepsilon_{ij}} \times \delta^\omega_{u_i}$;
9:    **while** $\vec{x}_{i+1} \neq \vec{x}_i \mathbf{p}$ **do**
10:       $\vec{x}_{i+1} = \vec{x}_i \mathbf{p}$;
11:    **end while**(The steady-state probability distribution vector $\vec{x}_{i+1} = (b^{\Theta(\tau)}_{u_i u_j})'$ is obtained by
      iteration.)
12:    Choose the maximum value of the steady-state probability distribution vector as the result
      of game selection. Add the node to the information diffusion process map: $SN' \Leftarrow Max\{f|f :$
      $(b^{\Theta(\tau)}_{u_i u_j}) \rightarrow (b^{\Theta(\tau)}_{u_i u_j})'\}$
13: **end for**
  Note: $'\leftarrow'$ and $'\Leftarrow'$ symbols in the algorithm represent assignment and adding values,respectively.

According to formula (8), we can calculate the OT-Page Rank value of each user in the process of information diffusion and obtain the final result $(b^{\Theta(\tau)}_{u_i u_j})'$ through iterated computing OT-Page Rank mapping function. Consequently, the maximum value in steady-state distribution vector is chosen as the result of user game selection.

Figure 1 shows an example of game selection in microblog social network, in which user u_1 post a message. According to proposed game selection rules, u_1 spreads the message to u_3. Then, the similarity of interest features between u_3 and his neighbors u_6, u_7 and u_8 is calculated, respectively. As a result, we take the values $(5/6, 5/9, 5/18)$ about the opportunity for user u_6, u_7 and u_8 to receive the message, respectively. The social trust of u_3 $(3/5)$ and the opportunities about all neighbors u_6, u_7 and u_8 are combined as the initial state probability distribution vector $(b^{\Theta(\tau)}_{u_3 u_j}) = \delta^\omega_{u_3} \times \varphi^\omega_{\varepsilon_{3j}} = \vec{x}_0 = (1/2, 1/3, 1/6)$ by OT-Page Rank.

The probability matrix $\mathbf{P}$ of the random walk process about node users is calculated according to the formula (8).

$$\mathbf{P} = \begin{bmatrix} 1/6 & 2/3 & 1/6 \\ 5/12 & 1/6 & 5/12 \\ 1/6 & 2/3 & 1/6 \end{bmatrix}$$

Then, the probability distribution after one step is:

$$\vec{x}_1 = (1/4 \;\; 1/2 \;\; 1/4) = \vec{x}_0 \mathbf{P}$$

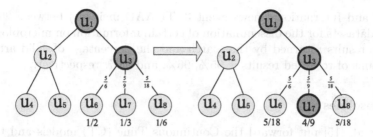

Fig. 1. An example of game selection based on OT-Page Rank algorithm.

The probability distribution after two step is:

$$\vec{x}_2 = \vec{x}_1 P = (1/4 \ 1/2 \ 1/4) \begin{bmatrix} 1/6 & 2/3 & 1/6 \\ 5/12 & 1/6 & 5/12 \\ 1/6 & 2/3 & 1/6 \end{bmatrix} = (7/24 \ 5/12 \ 7/24)$$

Table 2. A probability distribution vector sequence.

Probability vector	Probability distribution vector sequence		
$\vec{x}_0$	1/2	1/3	1/6
$\vec{x}_1$	1/4	1/2	1/4
$\vec{x}_2$	7/24	5/12	1/2
$\vec{x}_3$	13/48	11/24	13/48
$\vec{x}_4$	9/32	7/16	9/32
...	...	...	...
$\vec{x}$	5/18	4/9	5/18

The probability distribution vector sequence shown in Table 2 can be obtained by repeat iteration in the same way. The distribution is converged to the steady state $\vec{x} = (5/18 \ 4/9 \ 5/18)$. Then, we choose the maximum value 4/9 as the maximum benefit. Accordingly, user u_3 should select u_7 as the next receiver. The details of game selection information diffusion algorithm based on opportunity and trust is shown in Algorithm 1.

5 Experiment and Result Analysis

5.1 Data Collection and Analysis

Using the web crawler mechanism, this paper has crawled from Sina microblog (http://m.sinna.com.cn) to collect data from three microblog hot events (Social event 1: Chengdu female driver was beaten, Entertainment event 2: Baihe Bai derailed and International news event 3: THAAD incident) between 2015 and 2017 as datasets for the dissemination of certain information on microblog. Evaluate the results returned by the search and the percentage of valid articles in the amount of returned results is 95%, 96%, and 98%, respectively.

5.2 Baselines

Goyal et al. [15] put forward the Continuous Time (CT) models and the Discrete Time (DT) models to combine the general threshold models that predict the information propagation process. We use CT model in combination with the general threshold model as the comparison method since CT models have recognized the wisdom of finial prediction results compared with DT model. However, CT only calculates the probability of information propagation, which is called social influence. CT does not consider the structure influence factors from the perspective of network topology. Therefore, Li et al. [16] combined two kinds of global influences: influence of network structure and influence of information propagation, who proposed a Game Theory (GT) information propagation model. Consequently, we use the following three methods as baselines.

- CT + General threshold model: baseline method combining the CT model [15] with the general threshold model.
- GT (Page rank): GT model [16] use Page rank to calculate user's influence and use the social payoff to diffuse.
- GT (Propagation cascades): GT model use propagation cascades to calculate user's influence and use the social payoff to diffuse.

5.3 Experimental Results and Analysis

Three evaluative indexes of Precision, Recall and F1-Measure are used to evaluate the effect of the proposed model. This paper compare the result of the three models with proposed OT-Page Rank model based on opportunity, trust, and motivation. Experimental results are shown in Fig. 2.

Figure 2 shows the performance of focused approaches with different metrics at 7 time steps (One time step for 3 day) for the Sina microblog, respectively. In Fig. 2, blue curves (CT + General threshold model) render the results of baseline method. The green and orange one separately present the results of GT model combined with different influences on users' social payoff, respectively. The curves of green (GT (Page rank)) render the results of the GT model adopting Page rank values as user's influence. Orange one (GT (Propagation cascades)) render the results of the GT model adopting the property of diffusion cascades as user's influence. Our method is present with red curves that show the results of introducing evolutionary game theory into the OT-Page Rank model.

The experimental results show the effectiveness and superiority of the proposed method from three indexes. By adjusting the parameters continuously in the experiment, we get the optimal value of these parameters (see Table 3).

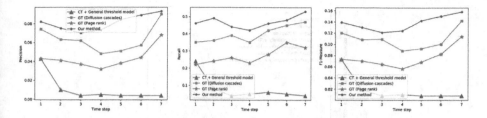

Fig. 2. Evaluative results of three different indexes based on Precision, Recall and F1-measure in Sina microblog. (Color figure online)

Table 3. The parameter values for events 1–3.

Parameter	Description	Event 1	Event 2	Event3
α_1	Sharing rate of users	0.658	0.536	0.598
α_2	Review rate of users	0.297	0.277	0.214
α_3	Praising rate of users	0.015	0.023	0.037
α_4	Users release posts' frequency within selected time	0.007	0.073	0.069
α_5	The weight of users release microblog posts	0.023	0.091	0.082
η_1	The weight of personal influence in social trust	0.245	0.144	0.236
η_2	The weight of content contribution in social trust	0.755	0.856	0.764

Figure 3 shows that after introducing different social factors and their combination factors, the rate of the number of information diffusion is different with the change of the series of information diffusion. Through simulation experiments, the proportion of the number of diffusion varies with the series of diffusion as showed by Num1, without any social factors. Num2, Num3 and Nmu4 respectively shows the trends of information diffusion after introducing three social factors: opportunity, trust and motivation. Num2 shows that users are more likely to get information about their interests or professions when people with similar interests spread information, which shortens the time they are accessing to the information. Num3 shows that users are more likely to generate trust with the information released by highly trusted users and accelerates the dissemination of information. Num4 shows that it is easier to make the propagation saturated when the user generates the motivation to propagate the information. Nums 5–8 shows the impact of the three social factors on the ratio of propagation in various combinations.

According to Fig. 3, we can see that different social factors have different perspectives and degrees of influence on information diffusion. Among them, opportunity can promote communication behavior and shorten the time for propagation contact, social trust accelerates the diffusion of specified information and shortens the life cycle of information diffusion, and motivation influences the depth and breadth of diffusion and makes important information as widely as possible for every individual in the microblog social network.

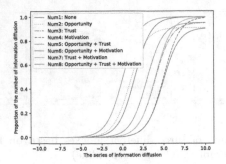

Fig. 3. Comparison of information diffusion ratio based on hybrid social factors.

6   Conclusion

This paper poses a highly efficient method to build a diffusion model by proposed Game selection based on Opportunity and Trust-Page Rank algorithm. Firstly, the similarity of interest between two adjacent users is creatively considered as the measure of opportunity factor that controls users to receive a certain message in the real world. Secondly, the features of special nodes with social influence and discourse power are defined as the threshold criterion on social trust. Finally, game selection contributes a lot to information diffusion by forming a key factor: motivation. The final model may plays an important role in helping researches model, analyze, and predict a more realistic information diffusion process.

Acknowledgement. Fruitful discussion with Yue Wu, Xianyong Li and Yongquan Fan is gratefully acknowledged. The authors thank anonymous reviewers for many useful discussions and insightful suggestions.

References

1. Zhang, Z., Wang, Z.: The data-driven null models for information dissemination tree in social networks. Physica A Stat. Mech. Appl. **484**, 394–411 (2017)
2. Zinoviev, D., Duong, V.: A game theoretical approach to broadcast information diffusion in social networks. In: Simulation Symposium, pp. 47–52 (2011)
3. Qiu, W., Wang, Y., Yu, J.: A game theoretical model of information dissemination in social network. In: International Conference on Complex Systems, pp. 1–6 (2013)
4. Jaccard, P.: The distribution of the flora in the alpine zone. New Phytol. **11**(2), 37–50 (1901)
5. Wu, S., Hofman, J.M., Mason, W.A., Watts, D.J.: Who says what to whom on Twitter. In: International Conference on World Wide Web, WWW 2011, Hyderabad, India, 28 March - April, pp. 705–714 (2011)
6. Kwak, H., Lee, C., Park, H., Moon, S.B.: What is Twitter, a social network or news media? In: International Conference on World Wide Web, pp. 591–600 (2010)
7. Daley, D.J., Kendall, D.G.: Epidemics and rumours. Nature **204**, 1118 (1964)
8. Pittel, B.: On a Daley-Kendall model of random rumours. J. Appl. Probab. **27**(1), 14–27 (1990)

9. Katz, E., Lazarsfeld, P.F.: Personal Influence: The Part Played by People in the Flow of Mass Communications. Transaction Publishers, Brunswick (2006)
10. Romero, D.M., Galuba, W., Asur, S., Huberman, B.A.: Influence and passivity in social media. In: International Conference Companion on World Wide Web, pp. 113–114 (2011)
11. Yong, S.K., Tran, V.L.: Assessing the ripple effects of online opinion leaders with trust and distrust metrics. Pergamon Press Inc, (2013)
12. Ortega, F.J., Troyano, J.A., Cruz, F.L., Vallejo, C.G., Enríquez, F.: Propagation of trust and distrust for the detection of trolls in a social network. Comput. Netw. **56**(12), 2884–2895 (2012)
13. Freeman, L.C.: Centrality in social networks conceptual clarification. Soc. Netw. **1**(3), 215–239 (1978)
14. Page, L: The PageRank citation ranking: bringing order to the web. In: Proceedings of the WWW Conference, pp. 1–14 (1998)
15. Goyal, A., Bonchi, F., Lakshmanan, L.V.S.: Learning influence probabilities in social networks, pp. 241–250 (2010)
16. Li, D., Zhang, S.P., Sun, X., Zhou, H.Y., Li, S., Li, X.L.: Modeling information diffusion over social networks for temporal dynamic prediction. IEEE Trans. Knowl. Data Eng. (99), 1 (2013)

Finding High-Quality Unstructured Submissions in General Crowdsourcing Tasks

Shanshan Lyu[1,2(✉)], Wentao Ouyang[1], Huawei Shen[1], and Xueqi Cheng[1]

[1] CAS Key Laboratory of Network Data Science and Technology,
Institute of Computing Technology, Chinese Academy of Sciences, Beijing, China
lvshanshan@software.ict.ac.cn
[2] University of Chinese Academy of Sciences, Beijing, China
{ouyangwt,shenhuawei,cxq}@ict.ac.cn

Abstract. The quality of crowdsourced work varies drastically from superior to inferior. As a consequence, the problem of automatically finding high-quality crowdsourced work is of great importance. A variety of aggregation methods have been proposed for multiple-choice tasks such as item labeling with *structured claims*. However, they do not apply to more general tasks, such as article writing and brand design, with *unstructured submissions* that cannot be aggregated. Recent work tackles this problem by asking another set of crowd workers to review and grade each submission, essentially transforming unstructured submissions into structured ratings that can be aggregated. Nevertheless, such an approach incurs unnecessary monetary cost and delay. In this paper, we address this problem by exploiting task requesters' historical feedback and directly modeling the submission quality, without the need of additional crowdsourced ratings. We first propose three sets of features, which try to characterize the submission quality from various perspectives, including the submissions themselves, the workers who make the submissions, and the interactions between task requesters and workers. We then propose two quality models, where one judges the submission quality independently and the other judges comparatively. These models not only incorporate features, but also take worker-specific factors into consideration. Experimental results on three large-scale data sets demonstrate that our models outperform general-purpose learning-to-rank methods such as Logistic Regression, RankBoost, and ListNet for finding high-quality crowdsourced submissions.

Keywords: Crowdsourcing · Quality estimation · General tasks
Unstructured submissions

1 Introduction

Crowdsourcing is becoming increasingly popular as it holds tremendous potential to solve a variety of problems that are difficult for computers in effective, efficient,

© Springer Nature Switzerland AG 2018
S. Zhang et al. (Eds.): CCIR 2018, LNCS 11168, pp. 198–210, 2018.
https://doi.org/10.1007/978-3-030-01012-6_16

and novel ways. Example tasks include image labeling [14], translation [19], logo design [1], and even software development [13]. Amazon Mechanical Turk[1], CrowdFlower[2] and Zhubajie[3] are examples of popular crowdsourcing platforms. Crowdsourcing offers remarkable opportunities for improving the productivity and the global economy by engaging a geographically distributed workforce to complete diverse tasks on demand and at scale. However, the quality of tasks completed by crowd workers is often lower than that by traditional employees and experts [7,13]. The problem of finding high-quality crowdsourced work is thus of great importance.

A variety of methods have been proposed for finding high-quality structured claims in multiple-choice tasks. For example, in an image labeling task where three workers claim the image labels as {"cat", "cat", "dog"}, one can employ majority voting to infer the correct claim as "cat". More sophisticated methods have also been proposed, considering aspects such as worker ability [5], task difficulty [17], worker community [16], and quantitative versus categorical claims [10,11]. Nevertheless, these methods work for structured claims and do not apply to more general tasks with diverse and unstructured submissions, such as article writing and brand design. Recent work tackles this problem by asking another set of crowd workers to review. However, crowdsourcing additional review incurs unnecessary monetary cost and delay.

In this paper, we address the problem of finding high-quality unstructured submissions by exploiting task requesters' historical feedback and directly modeling the submission quality, without the need of additional crowdsourced reviews. In crowdsourcing systems, workers make submissions to solve requesters' tasks, and requesters select high-quality submissions from workers. Intuitively, whether a submission has high quality that satisfies the requesters' need depends on many aspects, e.g., the properties associated with each submission, the properties of the worker who makes the submission, and the interactions between the task requester and the worker. Therefore, we extract three types of features: submission features, worker features, and requester-worker interaction features.

Given these features (as input) and requester feedback (as output), one possibility is to train popular learning-to-rank methods such as Logistic Regression [2], RankBoost [6], and ListNet [4] to rank submissions. However, these methods are general and are not specifically designed for crowdsourcing. To this end, we propose two more tailored quality models, namely, the *individual* and the *competition* quality models, where the former judges the submission quality independently while the latter judges comparatively. These models not only incorporate the extracted features, but also consider the worker-specific factors. Experimental results on three large-scale crowdsourcing data sets demonstrate that our models outperform popular general-purpose learning-to-rank methods.

In summary, our main contributions are as follows.

---

[1] https://www.mturk.com/mturk/welcome.
[2] http://www.crowdflower.com/.
[3] http://www.zbj.com/.

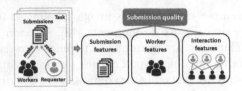

Fig. 1. Conceptual view of crowdsourcing (left) and the feature space that we consider for modeling submission quality (right).

1. We address the problem of finding high-quality unstructured submissions in general crowdsourcing tasks, in contrast to most existing studies on structured claims in multiple-choice tasks.
2. We propose three sets of features to capture various aspects from the submissions, the workers, and the requester-worker interactions, which are potentially correlated with the submission quality.
3. We propose two models, namely, the individual and the competition quality models, for estimating the submission quality. These models integrate both features and worker-specific factors in a unified framework. They do not need additional crowdsourced reviews.
4. We collect three large-scale real-world data sets from a leading crowdsourcing platform over a whole year. We conduct extensive experiments to evaluate the performance of different features and compare our proposed models with several popular learning-to-rank methods. We also make our data sets and implementation codes publicly available[4].

2   Problem Statement

We term the problem of finding high-quality crowdsourced submissions as the *quality estimation* problem for short. In this problem, the input is a set of submissions $\mathcal{S}_t = \{s_{t,i}\}_i$ for a crowdsourcing task t, and the output is the estimated quality $\hat{c}_{t,i}$ for each submission $s_{t,i}$. The estimated quality $\hat{c}_{t,i}$ can be used to re-rank the submissions in each new task in the order of descending quality, such that the requesters can identify high-quality submissions as soon as possible. The task t is characterized by the task requester r_t, task category, reward, requirements, posting time, and deadline. Each submission $s_{t,i}$ is characterized by the worker $u_{t,i}$, submission time, content (e.g., a logo), and additional text description. Requesters' feedback $c_{t,i}$ (i.e., selected or not) on the quality of past submissions is also available. We list the notations used in this paper in Table 1.

3   Feature Extration

In this section, we propose three sets of features that are potentially correlated with the quality of crowdsourced submissions. As shown in Fig. 1, in crowdsourcing systems, workers make submissions to solve requesters' tasks, and requesters

---

[4] URL is not provided for the review purpose.

Table 1. Notations and meaning.

Observed variables			
T	Total number of tasks	U	Total number of workers
L	Total number of features	t	Task identity
r_t	Requester identity of task t	$s_{t,i}$	The ith submission in task t
S_t	$S_t = \{s_{t,i}\}_i$; set of all the submissions in task t	N_t	Total number of submissions in task t
$c_{t,i}$	Feedback on the quality of $s_{t,i}$ by r_t; $c_{t,i} = 1$ if r_t selects $s_{t,i}$ and $c_{t,i} = 0$ otherwise	$\mathbf{x}_{t,i}$	Feature vector for submission $s_{t,i}$; $\mathbf{x}_{t,i} \in \mathcal{R}^{L \times 1}$
$u_{t,i}$	Worker identity of submission $s_{t,i}$		
Unknown variables to be inferred			
$\mathbf{w}$	Weight vector for features; $\mathbf{w} \in \mathcal{R}^{L \times 1}$	b	Global bias for features
h_u	Worker-specific parameter of u	$\hat{c}_{t,i}$	Estimated quality of a new submission $s_{t,i}$
θ	$\theta = \{\mathbf{w}, b, \{h_u\}_u\}$; set of model parameters		

select high-quality submissions from workers. Intuitively, whether a submission has high quality that satisfies the requesters' need depends on many aspects, e.g., the properties associated with each submission, the properties of the worker who makes the submission, and the interactions between the task requester and the worker. Therefore, we extract three types of features: submission features, worker features, and interaction features (Table 2).

In particular, we introduce how we extract the interaction features. There are two types of interactions: *participation* and *selection*. In the former, a worker (voluntarily) participates in a requester's task; in the latter, a requester selects a worker's submission. Clearly, only the selection behavior is quality-related. Therefore, we create a bipartite selection graph based on such behavior. We denote the selection graph as $\mathcal{G}_s = (\mathcal{V}, \mathcal{E}_s)$, where the node set $\mathcal{V} = \{\mathcal{U}, \mathcal{R}\}$ contains two types of nodes, i.e., the set $\mathcal{U} = \{u_i\}$ of workers and the set $\mathcal{R} = \{r_j\}$ of requesters. $\mathcal{E}_s$ is the edge set. An edge $e_{ij} \in \mathcal{E}_s$ is created if requester r_j selected worker u_i's submission in the past. After graph creation, we extract the following features: (1) *Edge weight.* Given worker u_i and requester r_j, we define this feature as the number of times that r_j selected u_i's submissions in the past. (2) *Node degree.* Given worker u_i, we define this feature as the number of requesters r_j connected to u_i in the selection graph. (3) *Authority score.* We run the HITS [8] algorithm on the selection graph where a requester functions as a hub, and a worker functions as an authority. After convergence, we use the authority scores of workers as features. We do not use the hub score of the requester because it keeps the same for any submission, which would not help to find high-quality submissions.

Table 2. Features for submission s (made by worker u) in task t (posted by requester r).

Submission features			
s-dl	Submission delay (in hours) of s after t is posted	s-len	Length of the text description associated with s
s-url	Whether the text description of s contains URL	s-at	Whether s has an attachment
s-vn	Version number of s		
Worker features			
w-nss	Number of selected submissions of u	w-rss	Ratio of selected submissions of u
w-rep	System reputation of u	w-tr	Total reward earned by u
w-ar	Average reward earned by u in a single submission	w-mr	Maximum reward earned by u in a single submission
Interaction features			
i-ew	Edge weight between u and r in the selection graph	i-nd	Node degree of u in the selection graph
i-as	Authority score of u based on the selection graph		

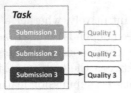 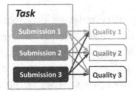

(a) Individual quality model   (b) Competition quality model

Fig. 2. Illustration of the individual and competition quality models. The quality of a submission is judged (a) based on itself, (b) with respect to all the submissions in the same task.

4   Submission Quality Models

In this section, we propose the *individual* and the *competition* quality models (Fig. 2), where the former judges the submission quality independently and the latter judges comparatively. These models not only incorporate the extracted features, but also considers the worker-specific factors.

4.1 Individual Quality Model

We first propose the individual quality model (IDV), where we model the quality of submission $s_{t,i}$ in task t as depending purely on itself. Mathematically, we model the probability $p_{t,i}$ that requester r_t selects submission $s_{t,i}$ as

$$p_{t,i} \triangleq p(c_{t,i} = 1 | s_{t,i}) = p(c_{t,i} = 1 | \mathbf{x}_{t,i}, \mathbf{w}, b, h_{u_{t,i}})$$
$$= 1 / (1 + \exp[-(\mathbf{w}^T \mathbf{x}_{t,i} + b + h_{u_{t,i}})]), \tag{1}$$

where $\mathbf{x}_{t,i}$ is the feature vector for submission $s_{t,i}$; $\mathbf{w}$, b, and $h_{u_{t,i}}$ are model parameters that to be inferred. These model parameters can be classified into two categories: (1) **global** parameters $\mathbf{w}$ and b, and (2) **local** parameters $h_{u_{t,i}}$. In particular, $\mathbf{w}$ is the global feature weight and b is the global feature bias. They relate the feature vector $\mathbf{x}_{t,i}$ to the submission quality through $\mathbf{w}^T \mathbf{x}_{t,i} + b$. Nevertheless, as $\mathbf{w}$ and b are global, they do not capture worker-specific variations in the data. Therefore, we introduce a worker-specific parameter $h_{u_{t,i}}$ for each $u_{t,i}$. The IDV model thus not only incorporate the extracted features, but also considers the worker-specific factors.

To alleviate the data sparsity problem, we impose a prior probability p_u on each worker-specific parameter h_u by $p_u \triangleq p(h_u | \sigma) = \mathcal{N}(h_u | 0, \sigma^2)$ where σ is a hyperparameter. The effect of $\ln p_u$ is equivalent to ℓ_2-norm regularization on h_u [2] (cf. Eq. (2)). The probability of observing the requester feedback $\{c_{t,i}\}_{t,i}$ and the worker-specific parameters $\{h_u\}_u$ in the IDV model is given by

$$P_{\text{IDV}} = \prod_{t=1}^{T} \prod_{i=1}^{N_t} \left[p_{t,i}^{c_{t,i}} (1 - p_{t,i})^{1-c_{t,i}} \right] \prod_{u=1}^{U} p_u,$$

where N_t is the total number of submissions in task t, T is the total number of tasks, and U is the total number of workers. To infer model parameters $\boldsymbol{\theta} = \{\mathbf{w}, b, \{h_u\}_u\}$, we maximize the log data likelihood, which is given by

$$\ln P_{\text{IDV}} = \sum_{t=1}^{T} \sum_{i=1}^{N_t} [c_{t,i} \ln p_{t,i} + (1 - c_{t,i}) \ln(1 - p_{t,i})] + \sum_{u=1}^{U} \ln p_u$$

$$= -\sum_{t=1}^{T} \sum_{i=1}^{N_t} \left[\ln \left[1 + \exp \left(-(\mathbf{w}^T \mathbf{x}_{t,i} + b + h_{u_{t,i}}) \right) \right] \right.$$

$$\left. + (1 - c_{t,i})(\mathbf{w}^T \mathbf{x}_{t,i} + b + h_{u_{t,i}}) \right] - \frac{1}{2\sigma^2} \sum_{u=1}^{U} h_u^2 - \text{const.} \tag{2}$$

4.2 Competition Quality Model

In the IDV model, each submission is judged individually, as either of high-quality or not. However, since the task reward is limited, even if many submissions are of high-quality when judged individually, the task requester cannot pick out all of them, but needs to find the top few best submissions. Therefore, we further propose the competition quality model (CPT), where submissions for the same task are considered to be competing and the submission quality is modeled comparatively. Mathematically, we model the probability $\tilde{p}_{t,i}$ of observing $c_{t,i} = 1$ given all the submissions $\mathcal{S}_t = \{s_{t,i}\}_i$ in task t as

$$
\begin{aligned}
\tilde{p}_{t,i} &\triangleq p(c_{t,i} = 1 | \mathcal{S}_t) = p(c_{t,i} = 1 | \{\mathbf{x}_{t,n}\}_n, \mathbf{w}, b, \{h_{u_{t,n}}\}_n) \\
&= \exp(\mathbf{w}^T \mathbf{x}_{t,i} + b + h_{u_{t,i}}) \Big/ \sum_{n=1}^{N_t} \exp(\mathbf{w}^T \mathbf{x}_{t,n} + b + h_{u_{t,n}}).
\end{aligned} \tag{3}
$$

where the probability of observing $c_{t,i} = 1$ now depends on all the submissions $\mathcal{S}_t$ in task t (reflected in its denominator), rather than only on the submission $s_{t,i}$ as that in the IDV model. No matter how large or small $\exp(\mathbf{w}^T \mathbf{x}_{t,i} + b + h_{u_{t,i}})$ is, as long as $\exp(\mathbf{w}^T \mathbf{x}_{t,i} + b + h_{u_{t,i}}) \gg \exp(\mathbf{w}^T \mathbf{x}_{t,n} + b + h_{u_{t,n}})$ for all $n \neq i$, we have $\tilde{p}_{t,i} \approx 1$ according to Eq. (3). In this way, CPT introduces comparison among submissions in the same task. Similar to the IDV model, the CPT model also integrates both features and worker-specific factors.

Since multiple submissions may be selected, we normalize the requester feedback $c_{t,i}$ for task t as $\tilde{c}_{t,i} \triangleq \frac{c_{t,i}}{\sum_{n=1}^{N_t} c_{t,n}}$. In this way, $\{\tilde{c}_{t,i}\}_i$ can serve as a valid empirical distribution since $\sum_i \tilde{c}_{t,i} = 1$. We also impose a prior probability on worker-specific parameter h_u as that in IDV model.

The probability of observing the normalized requester feedback $\{\tilde{c}_{t,i}\}_{t,i}$ and the worker-specific parameters $\{h_u\}_u$ in the CPT model is given by

$$
P_{\text{CPT}} = \prod_{t=1}^{T} \prod_{i=1}^{N_t} \tilde{p}_{t,i}^{\tilde{c}_{t,i}} \prod_{u=1}^{U} p_u.
$$

To infer model parameters $\boldsymbol{\theta} = \{\mathbf{w}, b, \{h_u\}_u\}$, we maximize the log likelihood:

$$
\begin{aligned}
\ln P_{\text{CPT}} = \sum_{t=1}^{T} \sum_{i=1}^{N_t} \tilde{c}_{t,i} \ln \tilde{p}_{t,i} + \sum_{u=1}^{U} \ln p_u = -\sum_{t=1}^{T} \Bigg[\ln \sum_{i=1}^{N_t} \exp(\mathbf{w}^T \mathbf{x}_{t,i} + b + h_{u_{t,i}}) \\
- \sum_{i=1}^{N_t} \tilde{c}_{t,i}(\mathbf{w}^T \mathbf{x}_{t,i} + b + h_{u_{t,i}}) \Bigg] - \frac{1}{2\sigma^2} \sum_{u=1}^{U} h_u^2 - \text{const.}
\end{aligned} \tag{4}
$$

4.3   Model Learning and Inference

Parameter Learning. We use the stochastic gradient descent (SGD) algorithm to learn model parameters of the IDV and the CPT models. We iterate over the training set 200 times. The hyperparameters are set to small fixed positive numbers. As the submissions are modeled as independent in IDV, we update the model parameters after each single submission. As the submissions for the same task are modeled as correlated in CPT, we update the model parameters after each task containing multiple submissions. The initial learning rate is set to $\eta = 0.02$ and it is reduced after each iteration.

Quality Estimation. After parameter learning, given a new task t with new submissions $\{s_{t,i}\}_i$, the IDV and the CPT models estimate the quality of each submission $s_{t,i}$ by applying Eqs. (1) and (3) respectively. Submissions are then sorted in the order of descending quality. For workers that do not appear in the training data set, these two models use the median of the estimated worker-specific parameters h_u over all the training workers as these new workers' parameters.

5 Experimental Results

5.1 Experimental Setup

Datasets. Our experiments are conducted over three real-world crowdsourcing data sets collected over one year from Zhubajie (ZBJ), which is a leading and the largest crowdsourcing platform in China. We retained tasks which were successfully completed (i.e., at least one submission was selected) in 2015 and whose category was *Brand design*, *Naming service*, or *Article writing*. These categories typical crowdsourcing tasks with different skill requirements. We then kept tasks with at least 5 submissions, as tasks with too few submissions can be easily assessed by requesters. The basic statistics of the resulting data sets are listed in Table 3.

Table 3. Basic statistics of the three crowdsourcing data sets (subs. - submissions).

Data set	# tasks	# subs	# selected subs	# requesters	# workers
Brand design	68,895	1,523,182	69,612 (4.57%)	55,788	17,188
Naming service	10,149	558,676	10,425 (1.87%)	9,503	19,796
Article writing	2,876	138,943	3,151 (2.27%)	2,640	9,150

Baselines. We compare the following popular learning-to-rank methods for finding high-quality crowdsourced submissions. They are (1) Regression Tree (RegTree) [2], (2) Logistic Regression (LogReg) [2], (3) RankNet [3], (4) RankBoost [6], (5) AdaRank [18], and (6) ListNet [4]. These methods take the feature vector $\mathbf{x}_{t,i}$ as input and the requester feedback $c_{t,i}$ as output. After training, given a new task with multiple submissions, these methods predict the feedback score for each submission and rank the submissions in descending quality. We do not implement the methods in [1] because they require another set of crowd workers to review each submission, which are thus impractical for large-scale evaluation.

Metics. We use three metrics that are widely used in the ranking problems [9] for evaluation, namely MAP, MRR, and NDCG.

Settings. Crowdsourcing tasks in each data set are sorted by their posting time. For each data set, we randomly pick a starting task, and then take the subsequent 800 tasks for training and the next 200 tasks for testing. We repeat this process for 20 times and report the results based on all the testing tasks. Each task contains on average 22.1, 55.0, and 48.3 submissions for the brand design, naming service, and article writing data set respectively. We preprocess the features as follows: (1) To alleviate the problem of extremely large feature span, we transform some features to the log scale. These features include s-dl, s-len, w-nss, w-rep, w-tr, and i-nd. (2) We then normalize each feature to the range $[0, 1]$ using the min-max normalization.

5.2   Performance of Individual Features

Table 4 lists the MAP values of each individual feature for unsupervised quality ranking. We make several key observations. (1) The relative performance of individual features vary in different data sets. It implies that our proposed features have different importance in revealing submission quality in different task categories. It also implies that each task category has its specific properties in terms of submission quality. (2) The length of additional text description associated with each submission (s-len) is the most powerful feature in all the data sets, showing its general effectiveness. (3) The ratio and the number of selected submissions (w-rss and w-nss) perform better than the system reputation (w-rep) in all the data sets. This is because the system reputation is computed across task categories, and thus it is less effective than category-specific measures. (4) The performance of the node degree in the selection graph (i-nd) and the number of selected submissions (w-nss) are quite similar across data sets. We find that most requester-worker pairs appear only once in our data sets. This may be because workers voluntarily participate in requesters' tasks and the chance of co-occurrence of a requester and a worker is small. This makes the number of requesters who selected a worker's submission (i-nd) similar to the number of selected submissions of a worker (w-nss).

Table 4. MAP of individual features. $R.$ – Rank; $Ft.$ – Feature.

Brand design						Naming service						Article writing					
R.	Ft.	MAP	R.	Ft.	MAP	R.	Ft.	MAP	R.	Ft.	MAP	R.	Ft.	MAP	R.	Ft.	MAP
1	s-len	0.6957	8	w-ar	0.2404	1	s-len	0.3762	8	w-rep	0.1683	1	s-len	0.4366	8	s-dl	0.2815
2	s-dl	0.3535	9	w-tr	0.2364	2	w-rss	0.2647	9	s-dl	0.1679	2	w-nss	0.3391	9	i-as	0.2799
3	w-rss	0.3326	10	s-at	0.2350	3	i-nd	0.2168	10	i-ew	0.1675	3	i-nd	0.3390	10	w-ar	0.2709
4	w-nss	0.2504	11	i-as	0.2316	4	w-nss	0.2162	11	s-at	0.1661	4	w-rss	0.3208	11	i-ew	0.2459
5	i-nd	0.2499	12	s-url	0.2194	5	w-tr	0.2064	12	s-url	0.1637	5	s-at	0.3200	12	w-rep	0.2430
6	w-rep	0.2434	13	i-ew	0.2133	6	i-as	0.1923	13	w-ar	0.1413	6	w-tr	0.2888	13	s-url	0.2332
7	w-mr	0.2418	14	s-vn	0.2046	7	w-mr	0.1718	14	s-vn	0.1273	7	w-mr	0.2853	14	s-vn	0.2310

Table 5. MAP of different feature sets using LogReg. S – submission features; W – worker features; I – interaction features.

Brand design			Naming service			Article writing		
Rank	Ft. set	MAP	Rank	Ft. set	MAP	Rank	Ft. set	MAP
1	$S+W$	0.7292	1	$S+W$	0.4382	1	$S+W$	0.4906
2	$S+I$	0.7184	2	$S+W+I$	0.4165	2	$S+W+I$	0.4807
3	S	0.7066	3	$S+I$	0.3925	3	S	0.4765
4	$S+W+I$	0.6960	4	S	0.3793	4	$S+I$	0.4717
5	$W+I$	0.3061	5	W	0.2534	5	W	0.3267
6	W	0.3048	6	$W+I$	0.2350	6	$W+I$	0.3085
7	I	0.2246	7	I	0.1991	7	I	0.2991

5.3 Performance of Feature Sets

In order to exploit the union of individual features and test whether the submission quality can be better predicted by considering a composition of signals. We test the performance of the submission feature set (S), worker feature set (W), interaction feature set (I), and their combinations using LogReg. We list the MAP values in Table 5. Again, the relative performance of different feature sets vary in different data sets. It is observed that the submission feature set performs significantly better than the worker and the interaction feature sets. This is because the submission feature set captures the dynamics and diversity of submissions (even by the same worker) while the other two types of features do not. For example, no matter whether a submission is timely or has adequate text description, the worker and the interaction features remain the same as long as the worker and the requester are the same. It is also observed that the combination of feature sets can lead to further performance improvement. As $S+W$ performs best across data sets, we use the $S+W$ feature set in all the following experiments.

5.4 Performance of Different Methods

In this section, we evaluate the performance of different methods. Table 6 lists the MAP, MRR and NDCG values. It is observed that MAP and MRR values are quite similar. This is because there is only one selected submission in most tasks (over 95%). As a result, considering all the selected submissions (by MAP) and the first selected submissions (by MRR) do not differ much. Among the general learning-to-rank methods, RankBoost performs best on the brand design and the naming service data sets, while LogReg performs best on the article writing data set. Sophisticated list-wise methods such as AdaRank and ListNet tend to overfit, especially on the naming service data set. Our IDV model outperforms all the general learning-to-rank methods, which shows that it is beneficial to model personalized worker-specific parameters in addition to extracting features

Table 6. Performance of different methods. The best-performing method is marked in bold. The best-performing baseline is underscored. selected submission in most tasks.

	Brand design			Naming service			Article writing		
	MAP	MRR	NDCG	MAP	MRR	NDCG	MAP	MRR	NDCG
RegTree	0.6704	0.6710	0.5695	0.4490	0.4500	<u>0.3310</u>	0.4329	0.4363	0.2947
LogReg	0.7292	0.7295	0.5870	0.4382	0.4389	0.2785	<u>0.4906</u>	<u>0.4928</u>	<u>0.3185</u>
RankNet	0.6360	0.6364	0.4588	0.3638	0.3642	0.2100	0.3702	0.3721	0.2063
RankBoost	<u>0.7443</u>	<u>0.7445</u>	<u>0.5998</u>	<u>0.4713</u>	<u>0.4718</u>	0.3015	0.4893	0.4908	0.3180
AdaRank	0.4908	0.4911	0.3428	0.2903	0.2905	0.1628	0.4612	0.4618	0.2873
ListNet	0.6832	0.6835	0.5253	0.2884	0.2889	0.1413	0.4371	0.4392	0.2688
IDV	0.8198	0.8202	0.7168	0.5583	0.5587	0.4108	0.5459	0.5474	0.3845
CPT	**0.8306**	**0.8313**	**0.7356**	**0.5704**	**0.5708**	**0.4276**	**0.5650**	**0.5666**	**0.4068**

based on domain knowledge. It is also observed that CPT even outperforms IDV and it achieves the best performance on all the three data sets. It shows that the competition view better explains the data than the individual view, thus leading to better quality estimation.

6   Related Work

Crowdsourcing Applications. Crowd workers have been explored to perform various tasks, such as image classification, sentiment analysis, and object counting [11,13,14], or even more creative tasks such as translation, article writing, and logo design [1,13]. However, these useful applications may be impaired by unskilled or sloppy crowd workers who provide low-quality output. The problem of identifying high-quality crowdsourced work is thus of great importance.

Aggregating Crowdsourced Claims. A variety of methods have been proposed for finding high-quality crowdsourced claims in multiple-choice tasks. Most of these methods work by aggregating redundant but structured claims. Majority voting is one of such methods. More sophisticated methods have also been proposed, considering aspects such as worker ability [5], spatial events [12], and others [20]. Nevertheless, these methods do not apply to unstructured crowdsourced submissions that cannot be aggregated.

Finding High-Quality Crowdsourced Submissions. Recent work [1,15] tackles this problem by asking another set of workers to review and grade each submission into a set of structured ratings (e.g., 1 to 5). Then the aforementioned methods can be then applied to aggregate these structured ratings. Nevertheless, crowdsourcing additional review incurs unnecessary monetary cost and delay.

Our proposed models overcome such limitations as they do not require additional crowdsourced ratings. Instead, our models exploit task requesters' historical feedback and directly model the submission quality. In this sense, our models are more practical and applicable.

7   Conclusion

In this paper, we address the problem of finding high-quality unstructured submissions in general crowdsourcing tasks. We propose three sets of features, i.e., the submission, worker, and interaction features, to capture various aspects that are potentially correlated with the submission quality. We further propose the individual and the competition quality models to automatically estimate the submission quality. These models not only incorporate extracted features, but also take worker-specific factors into consideration through personalized parameters. Large-scale evaluation results demonstrate that combination of the submission feature set and the worker feature set is the most effective, the competition quality model outperforms the individual quality model, and these two models both outperform popular general-purpose learning-to-rank methods for finding high-quality crowdsourced submissions.

References

1. Baba, Y., Kashima, H.: Statistical quality estimation for general crowdsourcing tasks. In: KDD, pp. 554–562. ACM (2013)
2. Bishop, C.M., Nasrabadi, N.M.: Pattern Recognition and Machine Learning. Springer, New York (2006)
3. Burges, C., et al.: Learning to rank using gradient descent. In: ICML, pp. 89–96. ACM (2005)
4. Cao, Z., Qin, T., Liu, T.Y., Tsai, M.F., Li, H.: Learning to rank: from pairwise approach to listwise approach. In: ICML, pp. 129–136. ACM (2007)
5. Dawid, A.P., Skene, A.M.: Maximum likelihood estimation of observer error-rates using the EM algorithm. J. R. Stat. Society. Ser. C (Appl. Stat.) **28**, 20–28 (1979)
6. Freund, Y., Iyer, R., Schapire, R.E., Singer, Y.: An efficient boosting algorithm for combining preferences. J. Mach. Learn. Res. **4**(Nov), 933–969 (2003)
7. Kittur, A., Chi, E.H., Suh, B.: Crowdsourcing user studies with mechanical turk. In: CHI, pp. 453–456. ACM (2008)
8. Kleinberg, J.M.: Authoritative sources in a hyperlinked environment. J. ACM **46**(5), 604–632 (1999)
9. Liu, T.Y.: Learning to Rank for Information Retrieval. Springer, Heidelberg (2011). https://doi.org/10.1007/978-3-642-14267-3
10. Ouyang, R.W., Kaplan, L., Martin, P., Toniolo, A., Srivastava, M., Norman, T.J.: Debiasing crowdsourced quantitative characteristics in local businesses and services. In: IPSN, pp. 190–201. ACM (2015)
11. Ouyang, R.W., Kaplan, L.M., Toniolo, A., Srivastava, M., Norman, T.J.: Aggregating crowdsourced quantitative claims: additive and multiplicative models. IEEE Trans. Knowl. Data Eng. **28**(7), 1621–1634 (2016)
12. Ouyang, R.W., Srivastava, M., Toniolo, A., Norman, T.J.: Truth discovery in crowdsourced detection of spatial events. IEEE Trans. Knowl. Data Eng. **28**(4), 1047–1060 (2016)
13. Quinn, A.J., Bederson, B.B.: Human computation: a survey and taxonomy of a growing field. In: CHI, pp. 1403–1412. ACM (2011)
14. Raykar, V.C., Yu, S., Zhao, L.H., Valadez, G.H., Florin, C.: Learning from crowds. J. Mach. Learn. Res. **99**, 1297–1322 (2010)

15. Sunahase, T., Baba, Y., Kashima, H.: Pairwise hits: quality estimation from pairwise comparisons in creator-evaluator crowdsourcing process. In: AAAI, pp. 977–984 (2017)
16. Venanzi, M., Guiver, J., Kazai, G., Kohli, P., Shokouhi, M.: Community-based Bayesian aggregation models for crowdsourcing. In: WWW, pp. 155–164. ACM (2014)
17. Whitehill, J., Wu, T.F., Bergsma, J., Movellan, J.R., Ruvolo, P.L.: Whose vote should count more: optimal integration of labels from labelers of unknown expertise. In: NIPS, pp. 2035–2043 (2009)
18. Xu, J., Li, H.: Adarank: a boosting algorithm for information retrieval. In: SIGIR, pp. 391–398. ACM (2007)
19. Zaidan, O.F., Callison-Burch, C.: Crowdsourcing translation: professional quality from non-professionals. In: HLT, pp. 1220–1229. ACL (2011)
20. Zhuang, H., Parameswaran, A., Roth, D., Han, J.: Debiasing crowdsourced batches. In: KDD, pp. 1593–1602. ACM (2015)

Recommendation

A Hybrid Neural Network Model with Non-linear Factorization Machines for Collaborative Recommendation

Yu Liu[1], Weibin Guo[1(✉)], Dawei Zang[1], and Zongyin Li[2]

[1] School of information Science and Engineering,
East China University of Science and Technology, Shanghai 200237, China
gweibin@ecust.edu.cn, 947510947@qq.com
[2] Institute of Science and Technology Development,
East China University of Science and Technology, Shanghai 200237, China

Abstract. In recent years, deep learning models have proven able to learn effective representation in many applications. However, the exploration of deep learning on recommender systems are relatively little. Although some recent work has utilized deep learning models to make recommendation, they primarily employed it to learn abstract representation of auxiliary information and used matrix factorization to model the interactions between user and item features. Especially, the application of deep learning models to learn user-item interaction function is very new and there are few attempts to this direction. In this paper, we propose a novel model Non-Linear Factorization Machine (NLFM) for modelling user-item interaction function and a hybrid deep model named AE-NLFM for collaborative recommendation. NLFM leverages neural networks to learn non-linear feature interaction and is more expressive than FM [15]. Extensive experiments on three real-world datasets show that our proposed AE-NLFM significantly outperforms the state-of-the-art methods.

Keywords: Collaborative filtering · Neural networks · Deep learning
Factorization machines · Representation learning

1 Introduction

With the popularization of various intelligent terminals, huge users have generated a huge amount of information, and information overload has become an obstacle for users to quickly find the items of interest. As a useful information filtering tool, recommender systems have attracted more and more attention from multiple disciplines [20].

Matrix factorization (MF) is widely used in personalized recommender systems. MF projects both users and items into a shared latent factor space and models the user-item interaction by the inner product in that space [8]. However, the inner product, as a simple interaction function, limits the performance

© Springer Nature Switzerland AG 2018
S. Zhang et al. (Eds.): CCIR 2018, LNCS 11168, pp. 213–224, 2018.
https://doi.org/10.1007/978-3-030-01012-6_17

of MF [7]. As the expansion of MF, factorization machines (FM) [15] are generic modeling of feature interactions. Compared to matrix factorization that only models the interactions (second-order) of two entities, FM can model the interactions (first-order and second-order) of any number of entities [5,19]. Despite the effectiveness of FM for many predictive tasks and collaborative filtering, it is well-known that its representation ability can be limited by the linear modeling of feature interactions [7]. FM cannot model the complicated non-linear feature interactions for real-world datasets.

In this work, we improve FMs representation ability by modelling the non-linear feature interactions. We proposed a novel model named non-linear factorization machines (NLFM), which utilizes the non-linear activation functions in neural networks to capture non-linear relationship between features. If we apply NLFM directly to make recommendations, the predictive performance of recommender system will suffer greatly for the data sparsity. To address this problem, we proposed a hybrid deep model named AE-NLFM, which integrates NLFM with deep feature learning. We utilize SDAEs [16] to extract high-level feature representation of rating and side information, and feed them into NLFM to get more accurate predictive ratings. Main contributions of this work can be summarized as follows.

1. We propose a novel NLFM model to improve FMs representation ability by learning non-linear feature interactions under the deep neural network.
2. We present a hybrid deep architecture named AE-NLFM to capture implicit relationship between users and items. AE-NLFM utilizes SDAEs to extract high-level feature representation and NLFM to conduct predictive task for collaborative filtering respectively.
3. We conduct extensive experiments on three real-world datasets to demonstrate the effectiveness of NLFM and AE-NLFM. Experiment results show that our models can significantly advance the state-of-the-art.

2   Related Work

Our work is closely related to the following two research areas: Factorization machines, and Deep learning based collaborative filtering. We will discuss them separately in the following subsections.

2.1   Factorization Machines

FMs were presented for collaborative recommendation, and now are mainly used in many prediction task, such as click through rates (CTR) [5] and rating prediction task [6]. By specifying input feature vector, many specific factorization models, such as standard MF and SVD++, can be seen as special cases of FM [7].

As one of the most promising methods for prediction task, FMs has attracted intensive research interests and many variants have been proposed [6,9,13,19].

Juan *et al.* [9] proposed field-aware factorization machine (FFM), which models interactions between different fields by learning multiple embedding vectors for each feature. Pan *et al.* [13] presented a novel sparse factorization machines (SFM) model that introduces Laplace distribution to model the parameters. Recently, deep learning based methods have been integrated with traditional FM. The Attentional Factorization machines (AFM) [19] is proposed to discriminate the importance of each feature interactions. He *et al.* [6] introduced a novel neural factorization machine (NFM) that captures higher-order feature interactions with deep neural networks.

2.2   Deep Learning Based Collaborative Filtering

It is well-known that deep neural networks can extract high-level feature representation from low-level raw feature data and learn an arbitrary function from real-world data. Deep learning based methods have emerged as a powerful tool for collaborative recommendation.

Recently, autoencoders (AE) and convolution neural network (CNN) are widely applied into feature representation extraction from side and textual information. Kim *et al.* [10] proposed a context-aware model named convolutional matrix factorization (ConvMF), which integrates CNN into PMF. Wang *et al.* [17] proposed a hierarchical deep Bayesian model named collaborative deep learning (CDL), which tightly couples the SDAE and PMF. Unlike CDL that only considers the content information of items, Li *et al.* [11] devised a general DCF architecture that employs the marginalized denoising autoencoders [2] to extract feature representation of the users and items content information. In [4], the authors introduced a hybrid collaborative filtering model with additional stacked denoising autoencoders.

In recent years, Many researchers employed deep neural networks to capture the relationship between users and items and learn the user-item rating function. Wu *et al.* [18] proposed a novel Collaborative Denoising Auto-Encoders (CDAE) model that utilized DAE to capture implicit relationship between users and items. Cheng *et al.* [3] developed a hybrid neural networks called Wide & Deep for app recommendation, where the Wide component is a linear model and the Deep component is a feed-forward neural network. He *et al.* [7] formulated a hybrid deep framework named neural collaborative filtering (NCF), which seamlessly combines the linearity of MF and non-linearity of MLP. By specifying activation functions and MLP, He *et al.* shows that NCF can mimic traditional MF.

In fact, our AE-NLFM is a combination of the two types. We employ the SDAEs to extract high-level feature representation of rating and side information, and utilize the NLFM to capture non-linear interactions and learn the user-item rating function.

3 Non-linear Factorization Machines

3.1 Factorization Machines

As a supervised machine learning model that works with real valued input vectors $\mathbf{x} \in \mathbb{R}^n$, factorization machines (FM) were proposed for collaborative recommendation [14]. FM estimates the target by modelling first- and second-order interactions between features:

$$\hat{y}_{FM}(\mathbf{x}) = \omega_0 + \underbrace{\sum_{i=1}^{n} \omega_i x_i}_{\text{first-order}} + \underbrace{\sum_{i=1}^{n-1} \sum_{j=i+1}^{n} \left(\mathbf{v}_i^T \mathbf{v}_j\right) x_i x_j}_{\text{second-order}} \tag{1}$$

where ω_0 is global bias, ω_i model the first-order interactions (linear regression) of the i-th feature to the target. The $\mathbf{v}_i = [v_{i1}, v_{i2}, ..., v_{ik}]$ term denotes the embedding vector of the i-th feature. According to [15], we can efficiently compute $\hat{y}_{FM}(\mathbf{x})$ in linear time $O(kn)$:

$$\hat{y}_{FM}(\mathbf{x}) = \omega_0 + \underbrace{\sum_{i=1}^{n} \omega_i x_i}_{\text{first-order}} + \underbrace{\frac{1}{2} \sum_{k}^{l=1} \left[\left(\sum_{i=1}^{n} v_{ik} x_i \right)^2 - \sum_{i=1}^{n} v_{ik}^2 x_i^2 \right]}_{\text{second-order}} \tag{2}$$

FM overcomes the shortcomings of traditional regression models for sparse data prediction by modelling second-order feature interactions. However, modeling for first-order interactions and second-order interactions are both linear models. So FM cannot capture non-linear interactions in real-world data.

3.2 The NLFM Model

In order to capture non-linear interactions in raw data, we use deep neural networks to model non-linear interactions. Similar to standard factorization machine, NLFM is a generic machine learning approach with any real vector $\mathbf{x} \in \mathbb{R}^n$. NLFM is defined as

$$\hat{y}_{NLFM}(\mathbf{x}) = \omega_0 + f(\mathbf{x}) \tag{3}$$

where ω_0 is global bias similar to that for FM. The second term $f(\mathbf{x})$ simultaneously models first- and second-order interactions. Figure 1 illustrates the neural network architecture of the term $f(\mathbf{x})$. In what follows, we introduce the structure of $f(\mathbf{x})$ layer by layer.

Input Layer. The input layers are two identical feature vector $\mathbf{x} \in \mathbb{R}^n$.

Embedding Layer. $f(\mathbf{x})$ has two different embedding layers. The vectors $[\omega_1], [\omega_2], ..., [\omega_n]$ are first-order embedding vectors (one-dimensional vector) corresponding to $x_1, x_2, ..., x_n$. The vectors $\mathbf{v}_1, \mathbf{v}_2, ..., \mathbf{v}_n$ are second-order embedding vectors (k-dimensional vector) corresponding to $x_1, x_2, ..., x_n$, where $\mathbf{v}_i = [v_{i1}, v_{i2}, ..., v_{ik}]$.

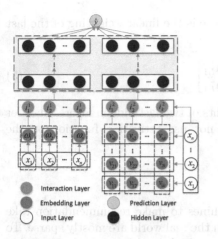

Fig. 1. Non-linear factorization machines (NLFM)

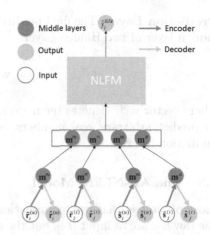

Fig. 2. The AE-NLFM model

Interaction Layer. $f(\mathbf{x})$ has two different interaction layers, whose input is first- and second-order embedding vectors and feature input. The first-order interaction of the input feature vector modeled by first-order interaction neurons $i_1^1, i_2^1, ..., i_n^1$, where $i_j^1 = \omega_j x_j$, $j = 1, 2, ..., n$. The vector formed by these first-order interaction neurons is called the first-order interaction vector, which is expressed as $\mathbf{I}^1 = \left[i_1^1, i_2^1, ..., i_n^1\right]$. The second-order interaction of the feature vector modeled by second-order interaction neurons $i_1^2, i_2^2, ..., i_k^2$, which are calculated by second-order embedding vectors and input feature vector:

$$i_l^2 = \left(\sum_{i=1}^n v_{il} x_l\right)^2 - \sum_{i=1}^n v_{il}^2 x_i^2, \quad l = 1, 2, ..., k \qquad (4)$$

Similar to first-order interaction vector , second-order interaction vector is expressed as $\mathbf{I}^2 = \left[i_1^2, i_2^2, ..., i_n^2\right]$.

Hidden Layer. $f(\mathbf{x})$ has two hidden layers, which are composed of fully connected layers:

$$\mathbf{h}_1^m = \sigma_1^m \left(\mathbf{W}_1^m \mathbf{I}^m + \mathbf{b}_1^m\right),$$
$$\mathbf{h}_2^m = \sigma_2^m \left(\mathbf{W}_2^m \mathbf{h}_1^m + \mathbf{b}_2^m\right), \qquad (5)$$
$$......$$
$$\mathbf{h}_{L_m}^m = \sigma_{L_m}^m \left(\mathbf{W}_{L_m}^m \mathbf{h}_{L_m-1}^m + \mathbf{b}_{L_m}^m\right)$$

Where m is equal to 1 or 2, indicating whether the hidden layer learns nonlinearity for first-order interactions or second-order interactions. The hidden layers takes the interaction vector $\mathbf{I}^1$ and $\mathbf{I}^2$ as inputs respectively. $\mathbf{W}_l^m$ and $\mathbf{b}_l^m$ denote the weight matrix and bias vector for the l-th hidden layer $\mathbf{h}_l^m$. The activation function of each hidden layer is non-linear activation functions, such as Rectifier (Relu) and Sigmoid, which is used to learn non-linear interactions.

Prediction Layer. The final prediction score is the linear weighting of the last hidden layer of two Hidden Layers.

$$\hat{y} = \mathbf{w}^{pre} \begin{bmatrix} \mathbf{h}_{L1}^1 \\ \mathbf{h}_{L1}^2 \end{bmatrix} \tag{5}$$

where vector $\mathbf{w}^{pre}$ denotes the neuron weights of two last hidden layers. In order to predict arbitrary real numbers, we do not use activation function on the prediction layer.

3.3   The AE-NLFM Model

Most of models that use factorization machines to make recommendation take the raw feature as input [14], but the data in the real world are mostly sparse. To address this problem, we use SDAE to extract high-level feature representation of raw feature. We then feed learned high-level feature representation into NLFM to provide more accurate recommendation.

Given the user-item rating matrix $\mathbf{R} \in \mathbb{R}^{m \times n}$, which m is the number of the users and n is the number of the items. We let $\mathbf{r}_i^{(u)}$ denotes the vector composed of i-th row of $\mathbf{R}$ and $\mathbf{r}_j^{(i)}$ denotes the vector composed of the j-th column of $\mathbf{R}$. $\mathbf{r}_i^{(u)} = [\mathbf{R}_{i1}, \mathbf{R}_{i2}, ..., \mathbf{R}_{in}]^T$ is a n-dimensional rating vector of user i on all the items and $\mathbf{r}_j^{(i)} = [\mathbf{R}_{j1}, \mathbf{R}_{j2}, ..., \mathbf{R}_{jm}]^T$ is a m-dimensional rating vector of item j rated by all the users. Numerous studies show that considering the auxiliary information can alleviate the cold start problem [10,17]. So we also use side information to improve performance. The vectors $\mathbf{s}_i^{(u)}$ and $\mathbf{s}_j^{(i)}$ denote the side information of the user i and item j respectively.

We first use four SDAEs to create high-level representations of rating and side information. We take the middle layers of four SDAE as high-level representations. The four middle layers are concatenated to serve as input to NLFM. Obviously, Fig. 2 illustrates the structure of AE-NLFM.

It indicates that the inputs of the hybrid neural network are sets $\tilde{\mathbf{r}}_i^{(u)}$, $\tilde{\mathbf{r}}_j^{(i)}$, $\tilde{\mathbf{s}}_i^{(u)}$ and $\tilde{\mathbf{s}}_j^{(i)}$. They are corrupted versions of rating and side information. AE-NLFM has five outputs $\hat{\mathbf{r}}_i^{(u)}$, $\hat{\mathbf{r}}_j^{(i)}$, $\hat{\mathbf{s}}_i^{(u)}$, $\hat{\mathbf{s}}_j^{(i)}$ and $\hat{y}_{ij}^{NLFM}$. The first four terms are reconstructions of rating and side information for user i on item j. The last term is predictive rating for user i on item j. The parameters of AE-NLFM are learned to minimize the reconstruction error and prediction error. Therefore, the objective function of AE-NLFM is formulated as follows:

$$L = \lambda_{pre} \sum_{i,j} \left(\mathbf{R}_{ij} - \hat{y}_{ij}^{NLFM} \right)^2 + \lambda_{rec} \cdot f + \lambda_{reg} \cdot g \tag{6}$$

where λ_{pre} and λ_{rec} are trade-off parameters, and λ_{reg} is a regularization parameter. It is worth noting that the value of $\lambda_{pre}/\lambda_{rec}$ has a huge impact on the performance of the model. The term f represent the reconstruction error of four SDAEs:

$$f = \sum_{i=1}^{m} \left(\mathbf{r}_i^{(u)} - \hat{\mathbf{r}}_i^{(u)} \right)^2 + \sum_{j=1}^{n} \left(\mathbf{r}_j^{(i)} - \hat{\mathbf{r}}_j^{(i)} \right)^2$$

$$+ \sum_{i=1}^{m} \left(\mathbf{s}_i^{(u)} - \hat{\mathbf{s}}_i^{(u)} \right)^2 + \sum_{j=1}^{n} \left(\mathbf{s}_j^{(i)} - \hat{\mathbf{s}}_j^{(i)} \right)^2 \tag{7}$$

And g denotes the regularization term for preventing overfitting:

$$g = \sum_{l} (\|\mathbf{W}_l^{ru}\|_F^2 + \|\mathbf{W}_l^{ri}\|_F^2 + \|\mathbf{W}_l^{su}\|_F^2 + \|\mathbf{W}_l^{si}\|_F^2$$

$$+ \|\mathbf{b}_l^{ru}\|_2^2 + \|\mathbf{b}_l^{ri}\|_2^2 + \|\mathbf{b}_l^{su}\|_2^2 + \|\mathbf{b}_l^{si}\|_2^2$$

$$+ \|\mathbf{W}_l^1\|_F^2 + \|\mathbf{W}_l^2\|_F^2 + \|\mathbf{b}_l^1\|_2^2 + \|\mathbf{b}_l^2\|_2^2) \tag{8}$$

where $\mathbf{W}_l^{ru}$, $\mathbf{W}_l^{ri}$, $\mathbf{W}_l^{su}$ and $\mathbf{W}_l^{si}$ denote the weight matrices for four SDAEs at layer l, and $\mathbf{b}_l^{ru}$, $\mathbf{b}_l^{ri}$, $\mathbf{b}_l^{su}$ and $\mathbf{b}_l^{si}$ are the corresponding bias vectors.

3.4   Model Learning

To optimize the objective function, we use stochastic gradient decent (SGD) to learn these parameters. Since most modern deep learning kits have provided the functionality of automatic differentiation, such as TensorFlow and Mxnet, we omit the derivatives here. Our models are implemented in TensorFlow and we employ mini-batch Adagrad as the optimizer. Additionally, to prevent overfitting, the dropout strategy has also been applied to hidden layers of four SDAEs and two hidden layers of NLFM.

4   Experiments

In this section, we perform extensive experiments on three real-world datasets from different domains to answer the following questions:

RQ1 How does our proposed hybrid neural network perform as compared to the state-of-the-art implicit collaborative filtering methods?

RQ2 Are four SDAEs helpful for AE-NLFM? How do the key hyper-parameters of SDAEs(the number of hidden layers) impact the performance of AE-NLFM?

RQ3 How do the value of $\lambda_{pre}/\lambda_{rec}$ impact the performance of AE-NLFM?

4.1   Experimental Settings

Datasets. In our experiments, we have selected the following three datasets to evaluate the effectiveness of our proposed models.

- **MovieLens-100k.** Similar to [4], we binarize the explicit data by regarding the instances with the ratings of four or higher as the positive instances. The positive instances have target value 1 and the negative instances have target value 0. We convert the user side information to a binary valued vector of length 879 via one-hot encoding. We convert the item side information to a binary valued vector of 260.

- **MovieLens-1M.** It contains 6040 users and 3706 movies with more than 1 million ratings, where each user has at least 20 ratings. Similar to MovieLens-100k, the user side information is encoded into a binary valued vector of length 30 and the item side information is encoded into a binary valued vector of length 99.
- **Frappe.**[1] [1] The frappe dataset consists of 96203 entries by 957 users for 4082 apps used in various contexts. As the frappe dataset contains positive instances only, we randomly sampled two negative instances for each positive instance. We convert all side information into a binary valued vector of length 343 using one-hot encoding and we only use three SDAEs to extract high-level representations in AE-NLFM.

Evaluation Protocol. To evaluate the prediction performance of all algorithms, we employ the root mean square error (RMSE) as the evaluation metric, where a lower scores indicates a better performance. For all the compared algorithms, we randomly sample instances from raw data at different percentages (70%, 80% and 90%) as the training set and the rest are used as test set. For each percentage, we conduct the evaluation five times for different randomly selected training and test set and the average value of RMSE is reported.

Baselines and Parameter Settings. In order to evaluate the performance of NLFM and AE-NLFM, we compare them with the following recommendation algorithms:

- **PMF.** [12] Probabilistic Matrix Factorization models latent factors of users and items by assuming there exists Gaussian observation noise and Guassian priors on the latent factors.
- **CDL.** [17] Collaborative Deep Learning is a hierarchical deep Bayesian model, which tightly couples the Stacked Denoising Autoencoders and PMF.
- **DCF.** [11] Deep Collaborative Filtering is a general framework for unifying deep learning approaches with collaborative filtering model.
- **NCF.** [7] Neural Collaborative Filtering is a framework aiming to capture the non-linear relationship between users and items.
- **NFM.** [6] Neural Factorization machines employ the non-linear neural network to capture high-order and non-linear feature interactions.

The batch size is set to 128 for all datasets. To be fair, we let the number of latent factors equal to the size of embedding vectors for all compared models. For CDL, DCF and our models, which are composed of DAE (or its variants), we test noise level of [0.0, 0.1, 0.2, 0.3, 0.4, 0.5] to get corrupted version of

---

[1] http://baltrunas.info/research-menu/frappe.

rating and side information. The regulation parameter λ_{reg} is set to 0.01 and the dropout ratio is set to 0.8 for our models. Without special mention, we only use one hidden layer to capture non-linear interactions.

4.2 Performance Comparsion (RQ1)

Table 1 show the average RMSE of different methods with respect to the size of embedding vectors. For MF based methods PMF, CDL and DCF, the number of latent factors is equal to the size of embedding vectors. From these results, we have the following observations.

Table 1. Average RMSE of compared models with different percentages of training data on three datasets

Dataset	Setting	k	PMF	CDL	DCF	NCF	NFM	NLFM	AE-NLFM
MovieLen 100k	70%	32	0.4846	0.4635	0.4539	0.4408	0.4369	0.4351	**0.4245**
		64	0.4756	0.4549	0.4471	0.4398	0.4338	0.4314	**0.4206**
	80%	32	0.4804	0.4588	0.4491	0.4359	0.4313	0.4287	**0.4187**
		64	0.4713	0.4530	0.4470	0.4351	0.4290	0.4249	**0.4145**
	90%	32	0.4783	0.4563	0.4475	0.4346	0.4282	0.4251	**0.4146**
		64	0.4695	0.4499	0.4452	0.4333	0.4255	0.4216	**0.4113**
MovieLen 1m	70%	32	0.4746	0.4585	0.4476	0.4273	0.4197	0.4193	**0.4103**
		64	0.4718	0.4533	0.4417	0.4252	0.4173	0.4163	**0.4086**
	80%	32	0.4718	0.4526	0.4401	0.4247	0.4170	0.4149	**0.4068**
		64	0.4626	0.4493	0.4366	0.4234	0.4147	0.4129	**0.4048**
	90%	32	0.4669	0.4495	0.4381	0.4231	0.4153	0.4124	**0.4043**
		64	0.4585	0.4467	0.4349	0.4219	0.4136	0.4109	**0.4029**
Frappe	70%	32	0.2146	0.2032	0.1958	0.1950	0.1828	0.1833	**0.1609**
		64	0.2007	0.1924	0.1867	0.1828	0.1661	0.1657	**0.1586**
	80%	32	0.2126	0.1987	0.1918	0.1907	0.1723	0.1713	**0.1524**
		64	0.1984	0.1883	0.1809	0.1778	0.1615	0.1606	**0.1504**
	90%	32	0.2121	0.1962	0.1887	0.1843	0.1669	0.1642	**0.1485**
		64	0.1963	0.1854	0.1801	0.1741	0.1592	0.1577	**0.1463**

First and foremost, we observe that AE-NLFM consistently achieves the best performance on all datasets. This demonstrates the effectiveness and rationality of SDAE in extracting high-level feature representation and NLFM in modelling non-linear feature interactions. Specially, AE-NLFM significantly outperforms the state-of-the-art methods NCF and NFM by a large margin on Frappe dataset.

Second, we find that AE-NLFM consistently outperforms NLFM on all datasets, which shows the effectiveness of SDAEs in capturing high-level presentation.

Lastly, PMF, CDL and DCF that are based on matrix factorization technique perform worse than NCF, NFM and our models that use neural networks to learn user-item scoring function. It demonstrates that the deep neural network is more effective than matrix factorization in modelling user-item scoring function.

4.3 Study of SDAEs (RQ2)

In this section, we focus on exploring the impact of SDAEs on the performance of AE-NLFM. We compare the performance of four models *w.r.t.* different sizes of embedding vectors. MLP-NLFM is the variants of AE-NLFM that ignores the decoder component of SDAEs. AE-2-NLFM indicates the AE-NLFM method with two hidden layers in SDAEs, and similar notations for others.

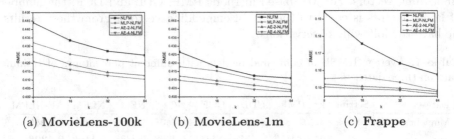

(a) **MovieLens-100k**   (b) **MovieLens-1m**   (c) **Frappe**

Fig. 3. Performance of RMSE *w.r.t.* the size of embedding vectors over three datasets

Figure 3 show the performance of four models *w.r.t.* different sizes of embedding vectors. As we can see, NLFM performs the worst on all datasets. It indicates that the high-level representation learned by MLP and SDAE is beneficial to performance. AE-4-NLFM and AE-2-NLFM achieve the best and second best performance, respectively. It demonstrates that SADEs can extract better and more abstract feature representation than MLP and stacking more layers on SDAEs are beneficial to performance.

4.4 Impact of the Ratio $\lambda_{pre}/\lambda_{rec}$ (RQ3)

As mentioned above, the value of $\lambda_{pre}/\lambda_{rec}$ has a huge impact on generalization performance of AE-NLFM. For convenience of analysis, we let λ_{rec} equal to 1. When λ_{pre} is extremely small, the value of $\lambda_{pre}/\lambda_{rec}$ will approach zero so that NLFM essentially vanishes. In this case our model will not optimize the prediction error, which measure the difference between the real rating and the predictive rating. Consequently, AE-NLFM degenerates to the unsupervised model SDAEs and cannot predict the rating of users on items. When λ_{pre} goes to positive infinity, the value of $\lambda_{pre}/\lambda_{rec}$ will approach positive infinity so that AE-NLFM degenerates to the supervised model MLP-NLFM, where the decoders of four SDAEs essentially vanishes.

Figures 4 and 5 show the training and test RMSE of AE-NLFM *w.r.t.* different value of $\lambda_{pre}/\lambda_{rec}$ on three datasets. The size of embedding vectors is set to 64 and the noise level is set to the optimal value for each dataset. As we can see, the greater the value of $\lambda_{pre}/\lambda_{rec}$, the faster the training RMSE will decline, which leads to a faster convergence for test RMSE. However, when the value of $\lambda_{pre}/\lambda_{rec}$ is greater than a certain value, although the training RMSE is reduced, the test RMSE does not decrease or even increases. Specifically, the optimal values of $\lambda_{pre}/\lambda_{rec}$ on three datasets are about 100 to 10000.

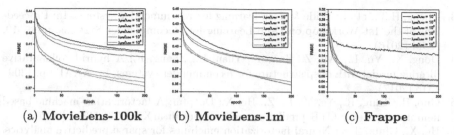

(a) **MovieLens-100k**   (b) **MovieLens-1m**   (c) **Frappe**

Fig. 4. Performance of training RMSE of each epoch $w.r.t.$ $\lambda_{pre}/\lambda_{rec}$ over three datasets

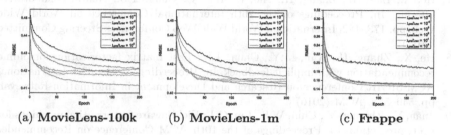

(a) **MovieLens-100k**   (b) **MovieLens-1m**   (c) **Frappe**

Fig. 5. Performance of test RMSE of each epoch $w.r.t.$ $\lambda_{pre}/\lambda_{rec}$ over three datasets

5 Conclusion and Future Work

In this paper, we explored deep learning model for collaborative recommendation. We presented a novel neural network model NLFM for modelling user-item interaction function and proposed a hybrid neural network model AE-NLFM for collaborative filtering. In fact, AE-NLFM provides a general neural framework, which can also admit deep learning models other than SDAE and NLFM. We can replace NLFM with AFM to discriminate the importance of feature interactions and replace SDAE with CNN to extract more informative textual and visual information. For further performance boost, we will investigate the other neural network model to find more accurate user-item interaction function and extract more informative feature representation.

Acknowledgements. This work is supported by the Natural Science Foundation of China (61672227, 61272198). The author would like to thank the anonymous reviewers for their reviewing efforts and valuable comments.

References

1. Baltrunas, L., Church, K., Karatzoglou, A., Oliver, N.: Frappe: Understanding the usage and perception of mobile app recommendations in-the-wild. arXiv preprint arXiv:1505.03014 (2015)
2. Chen, M., Xu, Z., Weinberger, K., Sha, F.: Marginalized denoising autoencoders for domain adaptation. Comput. Sci. (2012)

3. Cheng, H.T., et al.: Wide & deep learning for recommender systems. In: Proceedings of the 1st Workshop on Deep Learning for Recommender Systems, pp. 7–10. ACM (2016)
4. Dong, X., Yu, L., Wu, Z., Sun, Y., Yuan, L., Zhang, F.: A hybrid collaborative filtering model with deep structure for recommender systems. In: AAAI, pp. 1309–1315 (2017)
5. Guo, H., Tang, R., Ye, Y., Li, Z., He, X.: Deepfm: A factorization-machine based neural network for CTR prediction. arXiv preprint arXiv:1703.04247 (2017)
6. He, X., Chua, T.S.: Neural factorization machines for sparse predictive analytics. In: Proceedings of the 40th International ACM SIGIR Conference on Research and Development in Information Retrieval, pp. 355–364. ACM (2017)
7. He, X., Liao, L., Zhang, H., Nie, L., Hu, X., Chua, T.S.: Neural collaborative filtering. In: Proceedings of the 26th International Conference on World Wide Web, pp. 173–182. International World Wide Web Conferences Steering Committee (2017)
8. He, X., Zhang, H., Kan, M.Y., Chua, T.S.: Fast matrix factorization for online recommendation with implicit feedback. In: Proceedings of the 39th International ACM SIGIR Conference on Research and Development in Information Retrieval, pp. 549–558. ACM (2016)
9. Juan, Y., Zhuang, Y., Chin, W.S., Lin, C.J.: Field-aware factorization machines for CTR prediction. In: Proceedings of the 10th ACM Conference on Recommender Systems, pp. 43–50. ACM (2016)
10. Kim, D., Park, C., Oh, J., Lee, S., Yu, H.: Convolutional matrix factorization for document context-aware recommendation. In: Proceedings of the 10th ACM Conference on Recommender Systems, pp. 233–240. ACM (2016)
11. Li, S., Kawale, J., Fu, Y.: Deep collaborative filtering via marginalized denoising auto-encoder. In: Proceedings of the 24th ACM International on Conference on Information and Knowledge Management, pp. 811–820. ACM (2015)
12. Mnih, A., Salakhutdinov, R.R.: Probabilistic matrix factorization. In: Advances in Neural Information Processing Systems, pp. 1257–1264 (2008)
13. Pan, Z., Chen, E., Liu, Q., Xu, T., Ma, H., Lin, H.: Sparse factorization machines for click-through rate prediction. In: 2016 IEEE 16th International Conference on Data Mining (ICDM), pp. 400–409. IEEE (2016)
14. Rendle, S.: Factorization machines. In: 2010 IEEE 10th International Conference on Data Mining (ICDM), pp. 995–1000. IEEE (2010)
15. Rendle, S.: Factorization machines with libFM. ACM Trans. Intell. Syst. Technol. (TIST) 3(3), 57 (2012)
16. Vincent, P., Larochelle, H., Lajoie, I., Bengio, Y., Manzagol, P.A.: Stacked denoising autoencoders: learning useful representations in a deep network with a local denoising criterion. J. Mach. Learn. Res. 11(Dec), 3371–3408 (2010)
17. Wang, H., Wang, N., Yeung, D.Y.: Collaborative deep learning for recommender systems. In: Proceedings of the 21th ACM SIGKDD International Conference on Knowledge Discovery and Data Mining, pp. 1235–1244. ACM (2015)
18. Wu, Y., DuBois, C., Zheng, A.X., Ester, M.: Collaborative denoising auto-encoders for top-n recommender systems. In: Proceedings of the Ninth ACM International Conference on Web Search and Data Mining, pp. 153–162. ACM (2016)
19. Xiao, J., Ye, H., He, X., Zhang, H., Wu, F., Chua, T.S.: Attentional factorization machines: learning the weight of feature interactions via attention networks. arXiv preprint arXiv:1708.04617 (2017)
20. Zhang, S., Yao, L., Sun, A.: Deep learning based recommender system: A survey and new perspectives. arXiv preprint arXiv:1707.07435 (2017)

Identifying Price Sensitive Customers in E-commerce Platforms for Recommender Systems

Yingwai Shiu, Cheng Guo, Min Zhang$^{(\boxtimes)}$, Yiqun Liu, and Shaoping Ma

Department of Computer Science and Technology,
Beijing National Research Center for Information Science and Technology,
Tsinghua University, Beijing 100084, China
anson.tsinghua@gmail.com, gcmike@foxmail.com,
{z-m,yqunliu,msp}@tsinghua.edu.cn

Abstract. With the rise of E-commerce platforms, more people are getting used to make their daily purchases in online stores, especially for price-discounted goods. Therefore, online price cutting campaigns become a common approach for online retailers to compete with other competitors. Still, giving the same price discount to all may not be an efficient resource allocation as different users respond differently due to the differences of their price sensitivities. So if we are able to identify price sensitive users, both sellers and recommender systems will be greatly benefited from it in terms of improved user targeting and item suggestions. However, due to lack of detailed historical price and customer profile data, it is challenging to conduct price sensitivity analysis via traditional economics approach. More importantly, it is really hard and costly for companies to acquire price sensitivity labeled data. To overcome the constraints, making use of rich meta data (e.g. comment reviews) and time stamp becomes an alternative way. Inspired by distinct expressive power of graphical model, especially bipartite graph, we propose a User Behaviour Probability Transition Model (UBPT) which considers both user and item price sensitivities as weightings in the probability transition process. First, we define our own set of price sensitive users according to anonymous user after-purchase reviews. Second, we integrate selected behavioral features via doing user and item encoding. Third, using both user and item similarities, we combine our algorithm to simulate the probability transition process. With the data set from JD.com, our proposed model significantly outperforms other baselines in most cases. Besides, through applying the idea of UBPT to recommender systems, we can also enhance the performance of traditional recommendation algorithms.

Keywords: Price sensitivity · Recommender systems
Customer Identification

© Springer Nature Switzerland AG 2018
S. Zhang et al. (Eds.): CCIR 2018, LNCS 11168, pp. 225–236, 2018.
https://doi.org/10.1007/978-3-030-01012-6_18

1   Introduction

Different from traditional brick-and-mortar stores, within E-commerce platforms including Amazon, Taobao and JD.com, more frequent and significant price reductions are being adopted as a strategy to attract customers' attention and to stay ahead of fast-moving competitors. The huge success of Double Eleven Online Shopping Day, particularly in China, explains the popularity of online price discount campaigns towards customers.

Yet, giving the same price discount to all at a time may not be an effective resource allocation for suppliers, especially to small and medium ones with highly limited budget. It is because if sellers just simply lower the price without considering any customer attributes, especially towards price insensitive customers, sellers may suffer from lower-than-expected gain or even losses as shown in Fig. 1. Therefore, the ability to distinguish customers with different degree of price sensitivity is of paramount importance, in which its potential benefits are three fold: (1) It is useful for sellers to generate business intelligence in terms of better customer targeting. (2) It helps recommender system to suggest items based on price sensitivity and to increase customer satisfaction. [1] (3) It educates sellers on how to deal with their customers with different serving costs. For instance, E-commerce search engine can recommend items with price-unrelated benefits (e.g. more after-sales services) to the price insensitive customers while offering discounted items to attract price-sensitive users.

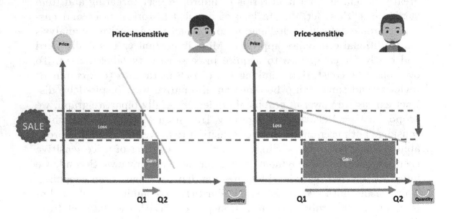

Fig. 1. For price insensitive users, the increase of quantity purchased will be less than the decrease of price under the same discount, hence bringing losses to the sellers.

To accomplish the task, several hurdles have to be overcome for identification of price-sensitive users. Firstly, in terms of data sufficiency, there is only limited pricing data, since comparing with non-internet market, there is always a smaller portion of customers who might be loyal to the same online shopping platform due to lack of human services and low switching cost [1], hence some previous works tend to create a simulated environment instead of detecting the actual relationship between a given price and the resulting revenue, not to mention understanding behavior of price insensitive customers. Secondly, as price sensitivity is a continuous variable, locating highly price-sensitive users who are likely to respond to price discount in the large extent is difficult. Thirdly, in terms of required data scope, if we want to consider all factors affecting price sensitivities, comprehensive customer profile and transaction logs from the core parts of the enterprise systems, are yet normally unavailable due to risk of leaking enterprise confidential business data and customer data privacy concerns. Fourth, most importantly, it is really costly to acquire price sensitivity labeled data due to its high complexity and scarcity.

To tackle the problems, we propose the User Behaviour Probability Transition Model (UBPT) that identifies groups of price sensitive users. Making use of two-year transaction logs from JD.com, one of the most popular E-commerce platform in China, we identify highly price-sensitive users by utilizing behavioral attributes from the transaction logs. Under the circumstance of missing pricing data and personal customer profile, we decide to make a full use of rich meta data, comment reviews, which is high likely to express customers intention on price sensitivity. Besides, we take the advantage of strong expressive power of graph model, and Bipartite graph model is chosen because of its propagating property. As the propagation depends on seed nodes which serve as the start of propagation, we introduce a set of seed users into the Bipartite graph. Since the user reviews involve the attitudes of users towards prices, we interpret different degree of price sensitivity based on percentage of expressed price consciousness in their comments. Then, starting from a small group of selected seed user data with the highest degree of price sensitivity, we would like to capture those who share similar shopping behavioral attributes as highly price sensitive customers. Besides, as an extension of our idea, we also use the same idea to enhance traditional recommendation algorithms which target on price sensitive customers. In sum, our major contributions of this paper are summarized as follows:

1. To the best of our knowledge, we are the first one who integrate graph model with user meta data in the task of identification of price sensitive customers.
2. We are able to use a small portion of labeled data to seek for price sensitive customers without using pricing, customer profile and price sensitivity labeled data.
3. We propose the User Behavior Probability Transition Model (UBPT), which can be universally and effectively applied to detect highly price sensitive customers in common E-commerce platforms.
4. By incorporating the concept of UBPT, we improve the traditional recommendation algorithm which focuses on price sensitive customers.

2   Related Work

In the field of economics and marketing, price sensitivity has been studied for years from traditional demand system to personalized promotion model. In definition, price elasticity of demand is defined as the sensitivity of demand (e.g.

online product sales) to price changes. [2] Traditionally, the idea of price sensitivity in recommender systems for E-commerce has been mentioned as a potential direction in the class survey. [3] On top of historical pricing data of the product, survey-level data that is available in the household level is being considered into the price sensitivity estimation model, in which in-depth consumer profile information is collected. Theoretically, such information would be insightful for retailers to more precisely identify the targeted customers, especially to whom with relatively higher degree of price sensitivity. Among existing works, price data has always been heavily used in order to compute price sensitivity numerically according to the economics definition of price elasticity formula. [5] For instance, people fed large transaction logs including product prices, meta-data and consumer profile information into the behavioral model to predict price sensitivity. [4] Price dynamics in multi-seller environment with price-sensitive customer has also been investigated via method of reinforcement learning. [5] Also, subjective pricing perception, such as reservation price (customer maximum willingness to pay) was also utilized for optimization of online promotion. [6] Furthermore, considering customer journey, one has attempted to define the price elasticity of demand in different stages of purchase by modeling customer preference and price sensitivity simultaneously by using large scale transaction log. [4] However, the very often cases require in-depth information about the enterprise system, including pricing data, customer personal profile, historical pricing, inventory stocks level in order to calculate numerical value of own-price elasticity and cross-price elasticity [2] etc., which are difficult to acquire at all times due to privacy concern. Due to the lack of labeled price-sensitive users, the problem is actually an unsupervised learning task, in which graph model is a favorable choice. For example, graph model can perform selection of textual labels for automatically generated topics [8], entity resolution [9] and estimation of similarity among objects [10] etc. Therefore, we are unique in a sense that (1) We apply a graphical probability model in the problem of estimation of price sensitivity by considering both user and item price sensitivity (2) We do not need price-label or price data throughout our estimation process; (3) We only require a small portion of clearly defined group of customers to identify a large group of price sensitive users; (4) Based on our observations, we simplify the methodology to acquire different degree of price sensitivity in an alternative way.

3   User Behavior Probability Transition UBPT Algorithm

3.1   Graph Representation of Price-Sensitive User

Within the bipartite model for the E-commerce scenario, there are two sets of nodes in opposite sides, which represent items and users respectively. Among all users and items, some of them are regarded as price sensitive users (Seeds). In the space between two sides of nodes, there are connections between user and item nodes, which represent at least one purchase action made by users. For the relationship between price-sensitive users, despite not directly linked together, they are indirectly connected by user-product-user bridges. In the most fundamental bipartite graph models, the seed nodes will propagate the information about the purchase action to other nodes according to the connections built by the purchase behavior. Also, in the propagation process, all connections $E = w_u i > 0, u, i \in G$, where E represents the set of connections and G stands for the set of nodes, have equal weights (Fig. 2). For the propagation process, the value is transfered by the factor of transition probability, which is shown below:

$$W_{ui} = W_{iu} = \frac{f_{ui}}{\sum_{u:(i,u) \subset E} f_{ui}} \tag{1}$$

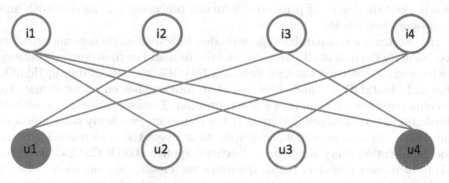

Fig. 2. For the annotation, The highlighted nodes are seed users, while u and i represent users and items respectively. u1 and u4 are connected with the u1-i3-u4 bridge.

3.2 User and Item Encoding

In order to better identify price sensitive users, on top of using the aforementioned graphical representation to explain the relationship between price sensitive users, we would like to better integrate both user and item features into consideration. Therefore we intend to use encoding method to embed all relevant features into vector representations. In our case, purchase time pattern and comment reviews are what we concern about.

(1) Purchase Time Pattern: In general, comparing with normal users, price sensitive users are more likely to make their purchases in the discount period. To have a better representation of discount periods, we carefully selected some major public holidays and famous Chinese online shopping festivals, in which majority of online retailers launch price-reduction campaigns to attract users. The discount periods we defined include National Day, Double 11, Double 12, Christmas, New Years Day and Mid-year Campaign 618. In addition, comparing with non-seed data, we also realized seed price sensitive users have a higher ratio of discount-day purchase to all purchases. Therefore we could treat discount period purchase as a comparable attribute to infer degree of price sensitivity.

(2) Comment Reviews: Under the circumstance that price and personal customer profile data are missing, we can no longer rely on traditional economics methods, which measure the ratio of percentage change of quantity demand to percentage change of price. As making a comment is not a compulsory step during the online purchase, assuming the crowd follows the law of lazy users, we tend to believe that the comments users gave reflect their genuine concern about the purchase. Besides, we also find out customer reviews contain rich information about the user attributes, including price sensitivity. Thus, applying similar methodologies to solve unavailable linkage variable in traditional price sensitivity prediction task, we can also make use of anonymous user after-purchase review for validation purpose. Our key assumption is that if a user explicitly leave a price-related comment after online purchase, in some extent he/she has shown a certain degree of price consciousness, especially for the case with high mentioning frequencies.

To construct a comparable representation for every single user and item, we need to unify the unit of all attributes into vector first. For time purchase pattern, we have extracted time features including DayOfWeek, DayOfMonth, HourOf-Day and MonthOfYear, and converted them into one hot encoding format. For discount period, we remain it as 1 or 0 expression. Then we take average for both attributes to the user-level information. For comments, we firstly stack user-level comments into separate small paragraphs, so as to get the most textual information representing every single user. Next, we applied the DOC2VEC algorithm [11] to turn every comment paragraph into a vector with 50 dimensions, which is similar to the average length of the comment paragraphs. During the conversion, we remove common stop words. For parameters, we set minimum word counts to give higher frequency words more weighting than the others. As a result, we will have a vector with 113 dimensions to describe every single user for further processing. Similarly, considering item representation, we need to do the same series of calculations for items. For selection of seed records, we interpret the items purchased by seed users as seed items. For the usage of features, we switch it to item's perspective, including the time being purchased, whether being purchased in the selected discount periods as well as all comments associated with every item.

3.3   Similarity-Based Transition

From user's point of view, one person may respond in contrast manner towards the same price discount in two different items due to different perceived price sensitivities of items. From item's point of view, two users may act differently towards same price discount on the same item due to different users' price sensitivities. Therefore, we think it is necessary to assign different weightings to different user-item pairs, hence different combinations of item and user price sensitivity can be explained by the algorithm. See Fig. 3.

In fact, there are more than three combinations of item and user pair as price sensitivity is measured in continuous scale. To assign different degree of price sensitivity to every user and item, we intend to express it in terms of similarity. For seed users, we interpret it as highly price sensitive user, so they have the highest price sensitivity, in which its value is set as 1. By comparing the cosine similarity between every single non-seed user/item and the average vector representation of seed (user/item), we are able to get its relative price similarity for all non-seeds in range from 0–1. In order to incorporate both item and user price sensitivity in each user-item pair, we combine cosine similarities as follows:

$$w_{ui} = w_{iu} = \lambda_1 sim_u + \lambda_2 sim_i + (1 - \lambda_1 - \lambda_2)sim_u * sim_i \qquad (2)$$

Therefore, after assigning different weightings in different user-item pairs, given seed item/user always reset to 1 in every iteration, user will be more likely to propagate more to the price sensitive items while item will also transfer more back to the price sensitive uses. Ultimately, the user with the higher sensitivity will get higher probability.

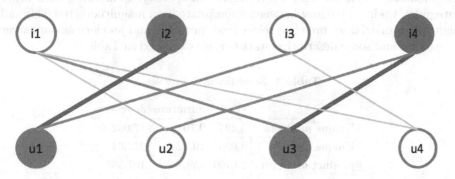

Fig. 3. The thickness of the line represents weighting in different user-item pairs. In the heaviest weighting pair, both user and item are seeds. For the second heaviest pair, either one user or item is seed. In the least heavy weighting pair, both user and item are non-seeds.

3.4   Text-Based Seeds Mining

In UBPT, seeds play a critical role in the propagation process, in our case, interpreted as highly price-sensitive users. To solve the problem of high labeling cost of price-sensitivity labeled data, we prefer using rich information in comment review, which high likely to include user's intension of price sensitivity. Given every user in our experiment has at least 5 reviews, we regard the users whom 100% of his/her comments containing any of the price-related keywords, including "expensive", "affordable", "high price-quality ratio", "Free Gift with Purchase", "Special Offers" and "High Shipping Cost" etc. as our seed users.

4   Experiments

4.1   Dataset

We have collected transaction logs from 1/9/2012–3/5/2014 with the help of
JD.com, which is one of the largest E-commerce platform in China. Within the
dataset, we have features stated as below:

In the data set, no private information (e.g. customer profile) or historical
selling price was included. Before further data-processing, we remove anonymous
users to ensure proper experiment operation. Also, to avoid extreme case for the
positive feedback problem brought by 1-degree nodes [7], we only keep users with
at least 5 purchases, which means 5 connections per node in the UBPT graphical
model at least. To have a clearer focus on users degree of price sensitivity, we
conduct our experiment on three separate categories, so as to investigate the
phenomenon of different price sensitivities on different item categories. Among
three level of categories, we have decided to work in the first level, which is
the broadest, to ensure sufficiently high data sparsity. Comparing with other
categories, taking a balance between appropriate data magnitude (100,000) and
high sparse matrix, we prefer to select food, furniture and jewelery as our target
categories, and some descriptive statistics are displayed in Table 1:

Table 1. Statistics of 3 data sets

	Food	Furniture	Jewellery
#Unique products	4,494	9,704	17,032
#Unique users	3,030	19,123	21,364
#product user pairs	23,100	139,786	167,556
#seed user	84	283	898

4.2   Baselines

As our experiment is a classification task to distinguish price-sensitive user from
other users, we select traditional classifiers shown as below:

SVM: It trains a frontier which best separates two classes of data

Decision Tree: A tree structure is constructed that breaks the dataset down
into small subsets. More important attribute will be prioritized to be decision
nodes, while partition the others into leaf nodes

KNN: The Nearest Neighbours method is to find a predefined number of training samples closest in distance to the new point, and predict the label from
these.

Random Forest: It builds many decision trees and combine them together to
a get a more accurate and stable prediction.

Naive Bayes: It is a family of simple probabilistic classifier based on Bayes
theorem with strong independence assumptions between nodes.

4.3   Results and Analysis

Table 2. Experiment result: In most datasets, our proposed model *bi-UBPT* is better than baseline models. On top of traditional classifier, *bi-origin* represents the original bipartite graph model, while *bi-price* represents the bipartite graph model with different weighting of probability transition, were the weight equals to the percentage of comments containing price related keywords

	Food			Furniture			Jewellery		
	f1	rec	pre	f1	rec	pre	f1	rec	pre
KNN	0.0526	0.0286	0.3316	0.0896	0.0490	0.5470	0.0580	0.0267	0.3766
Random forest	0.0538	0.0320	0.3730	0.0328	0.0172	0.3414	0.0088	0.0044	0.1620
Naive Bayes	0.1062	0.0892	0.1540	0.1248	0.1094	0.1466	0.0716	0.1232	0.0508
SVM	0.1394	0.0790	0.6538	0.1020	0.0528	0.5816	0.0410	0.0214	0.4410
Decision tree	0.1504	0.1322	0.1782	0.1334	0.0972	0.1970	0.0788	0.0544	0.1474
bi-origin	0.1838	0.5806	0.1164	0.1543	0.9412	0.0841	**0.0860**	0.9570	0.0451
bi-price	0.1550	0.3451	0.1130	0.1457	0.8077	0.0825	0.0853	0.9399	0.0447
bi-UBPT	**0.1967**	0.5922	0.1198	**0.1543**	0.9077	0.0847	0.0859	0.9444	0.0450

From the above Table 2, it shows the performance of highly price sensitive customer detection with different methods on three evaluation metrics, F1 score (f1), Recall (rec) and precision (pre).

As we can observe from the table, all models originated from bipartite graph significantly outperform other classification baseline methods, which showcases bipartite graph is an effective model in the classification task under the constraint of limited information and lack of price-sensitivity label about the users and items. When we compare among bi-origin, bi-UBPT and bi-price, our proposed model has a better performance in both Food Dataset, while perform more or less the same in the furniture and jewelery dataset. One of the possible reasons are due to different item nature and user's perceived price level. In item level, comparing with furniture and jewelry, food in general has a lower price and larger price variance. In user level, furniture and jewelry consume a larger proportion of total expenditure, which increases the price elasticity in general. Besides, this phenomenon also reflects the situation in our dataset, in which the portion of price-related comments in food category (ground truth: 14.26%) is higher than furniture (ground truth: 9.76%) and jewelry (ground-truth: 8.48%) categories. As a result, given users in general regard the purchase of furniture and jewelery in the online platform as expensive consumptions, they are more likely to leave comments about factors other than price (e.g. quality) [1] instead of normal expectation of high price. Thus, the behavior distinction between verbally price aware customers and non price aware customers in the online platform becomes less and the performance of bi-origin and our proposed model bi-UBPT will be indifferent. One point to note about bi-price: as bi-price perform the worst among three bipartite graph-based method, it implies percentage of price-related keywords is unsuitable for being the transition probability in the bipartite graph.

4.4 Training and Testing

For the division of dataset, to minimize the error brought by data distribution, we have conducted a five fold cross validation. For the training set, we randomly sample 80% of data as training set while keeping 20% of data as the hidden set. For the test set, we combine five different randomly sampled training sets with 5 randomly sampled hidden sets. To maximize the propagation effect of seed, we intentionally locate all seeds in the training set. Therefore the number of seed in training set and test set are the same. Also, as we perceive a price-sensitive user is high likely to frequently express price-related key words in their comments, users (excluding seed users) who express price-sensitive keywords for at least 50% of all his/her comments, in which we discovered only a small portion can fulfill the criteria, will be regarded as our ground truth.

In training set, our goal is to obtain an optimal threshold value and the number of rounds. In the test set, we will extract the output of a descending order list of ultimate probability for every user in the learnt number of rounds. (e.g. 54th rounds) Then, using the trained threshold as a classifier, only the ones' (excluding seed users) probability larger than the threshold will be categorized as price sensitive users and vice versa. At the end, we will have f1-score, precision and recall as our evaluation metrics.

5 How Price-Sensitivity Helps in Recommendations

As we mentioned in the beginning, one of the most important applications of price-sensitive user identification is to help improve recommendations pushed by E-commerce recommender systems. In this section, we will introduce how price-sensitivity-based similarity is used to give more accurate recommendations. Firstly, we take advantages of 5 widely used recommendation algorithms in modern online-shopping scenarios, including MostPOP, UserKNN [14], ItemKNN [13], WRMF [15] and BPRMF [12]. Then we investigate all these methods on the dataset of jewellery (one of the largest dataset). For each user in the dataset, we randomly select 80% of their purchasing records as training set, and the rest for test set. Then we select the top-50 recommendation results, together with the corresponding prediction scores from each method for further promotion. Intuitively, this promotion is to recommend price-sensitive products to price-sensitive users. As a result, we make use of the price-sensitive-based similarity (raised in Sect. 4) to measure whether a product matches a user. The final prediction score of recommending an item to a user is calculated as follow:

$$p_f = p_r * s_{ui} \tag{3}$$

where p_r is the prediction score from recommendation algorithms and s_{ui} is the price-sensitive-based similarity between user u and item i. We could then give a newly ranked recommendation list according to p_f to each user. Here we use RECALL as the evaluation metric to see whether our method to give a better ranking order. In Table 3, it shows the performance of different recommendation methods, where "before" means the ranking results from original methods and "after" means the results given after integrations with price-sensitivity-based similarity. From the result, we could conclude that BPRMF outperforms almost all other original methods, except perform similarly as MostPop. Besides, the improved performances of most these updated methods indicate that recommending price-sensitive products to price-sensitive users is reasonable and appropriate.

Table 3. Comparison between before and after integration of similarity

Recall (Top 50)	BPR	WRMF	ItemKNN	UserKNN	MostPop
Before	0.8291	0.6304	0.8202	0.8214	0.2318
After	0.8292	0.6317	0.8229	0.8240	0.2317

6 Conclusions and Future Work

In this paper, we have proposed a novel graphical probability model, User Behavior Probability Transition Algorithm (UBPT), in E-commerce recommender system to identify price sensitive users. Different from traditional economics methods and current approaches, we do not need to rely on historical price record or detailed customer profile. Essentially, to overcome the high acquisition cost of price sensitivity-labeled data, UBPT provides another possibility to leverage on rich meta data (comment reviews), which is commonly available in the E-commerce transaction log. Inspired by the structure of bipartite graph, since we realize it is necessary to consider both price sensitivity of users and items, therefore we assign differential weightings to different user-item pair probability transitions. From the experiment result, it shows that our algorithm outperforms other common baseline classification models. Furthermore, through applying the idea of UBPT on the traditional recommendation algorithm to push item suggestions to price sensitive users, enhanced performances are shown in our results.

For future work, several worthwhile tasks can be done. Firstly, building on UBPT, which shows signs of improvement over traditional classification models, continuous integration with deep learning models will also be explored. Secondly, more user experiments and analyses can be done in order to investigate more about the behavioral pattern of price sensitive users. Thirdly, next time we can also consider the effects of purchase time sequence and quantity in our model. Lastly, it is even ideal for us if we can practically test our algorithm in more different item categories in the operating E-commerce platform to test it effects.

236   Y. Shiu et al.

References

1. Mercy, O.: Price flexibility in relation to consumer purchasing behaviour on-line (business to consumer electronic commerce) (2009)
2. Ghose, A., Sundararajan, A.: Evaluating pricing strategy using E-commerce data: evidence and estimation challenges. Stat. Sci. 131–142 (2006)
3. Schafer, J.B., Konstan, J., Riedl, J.: Recommender systems in E-commerce. In: Proceedings of the 1st ACM Conference on Electronic Commerce, pp. 158–166. ACM, November 1999
4. Wan, M., et al.: Modeling consumer preferences and price sensitivities from large-scale grocery shopping transaction logs. In: Proceedings of the 26th International Conference on World Wide Web, pp. 1103–1112, April 2017
5. Chinthalapati, V.R., Yadati, N., Karumanchi, R.: Learning dynamic prices in multiseller electronic retail markets with price sensitive customers, stochastic demands, and inventory replenishments. IEEE Trans. Syst. Man Cybern. Part C (Appl. Rev.) 36(1), 92–106 (2006)
6. Musial, J., Pecero, J.E., Lopez, M.C., Fraire, H.J., Bouvry, P., Blazewicz, J.: How to efficiently solve internet shopping optimization problem with price sensitive discounts? In: 2014 11th International Conference on e-Business (ICE-B), pp. 209–215. IEEE, August 2014
7. Wei, C., Liu, Y., Zhang, M., Ma, S., Ru, L., Zhang, K.: Fighting against web spam: a novel propagation method based on click-through data. In: Proceedings of the 35th International ACM SIGIR Conference on Research and Development in Information Retrieval, pp. 395–404. ACM, August 2012
8. Aletras, N., Stevenson, M.: Labelling topics using unsupervised graph-based methods. In: Proceedings of the 52nd Annual Meeting of the Association for Computational Linguistics (Volume 2: Short Papers), vol. 2, pp. 631–636 (2014)
9. Zhu, L., Ghasemi-Gol, M., Szekely, P., Galstyan, A., Knoblock, C.A.: Unsupervised entity resolution on multi-type graphs. In: Groth, P., et al. (eds.) ISWC 2016. LNCS, vol. 9981, pp. 649–667. Springer, Cham (2016). https://doi.org/10.1007/978-3-319-46523-4_39
10. Muthukrishnan, P.: Unsupervised graph-based similarity learning using heterogeneous features. Doctoral dissertation, University of Michigan (2011)
11. Le, Q., Mikolov, T.: Distributed representations of sentences and documents. In: International Conference on Machine Learning, pp. 1188–1196, January 2014
12. Rendle, S., Freudenthaler, C., Gantner, Z., Schmidt-Thieme, L.: BPR: Bayesian personalized ranking from implicit feedback. In: Proceedings of AUAI, pp. 452–461, June 2009
13. Deshpande, M., Karypis, G.: Item-based top-n recommendation algorithms. ACM Trans. Inf. Syst. (TOIS) 22(1), 143–177 (2004)
14. Konstan, J.A., Miller, B.N., Maltz, D., Herlocker, J.L., Gordon, L.R., Riedl, J.: GroupLens: applying collaborative filtering to Usenet news. Commun. ACM 40(3), 77–87 (1997)
15. Hu, Y., Koren, Y., Volinsky, C.: Collaborative filtering for implicit feedback datasets. In: Eighth IEEE International Conference on Data Mining, ICDM 2008, pp. 263–272. IEEE, December 2008

Jointly Modeling User and Item Reviews by CNN for Multi-domain Recommendation

Yong Cai, Shoubin Dong(✉), and Jinlong Hu

Communication and Computer Network Key Laboratoty of Guangdong,
School of Computer Science and Engineering,
South China University of Technology, Guangzhou, China
eeyongcai@mail.scut.edu.cn, {sbdong, jlhu}@scut.edu.cn

Abstract. With the rapid development of e-commerce platforms, the change of shopping ways brings people more choices as well as the overload of the information when shopping. Recommender systems, as the main technology to solve information overload, have been applied to product recommendation successfully. However, the majority of the traditional recommender systems, which are based on the overall rating information of user and focus on single domain recommendation, couldn't take advantage of user review and the common consumption pattern among multi domains to achieve a better recommendation. In this paper, a multi domain product recommendation algorithm CCoNN based on Convolutional Neural Network (CNN) is proposed, which leverage the common user's review information among several domains to generate user preference vector and item features vector. After that, such vectors are used to make overall rating prediction for a user-product pair through a Factorization Machine (FM). Experiments show that the algorithm has a better performance than other reviews-based recommendation method.

Keywords: Multi-domain recommendation · Convolutional neural network
Reviews-based recommendation system

1 Introduction

With the rapid development of Web2.0, the e-commerce platform has gradually become a major shopping way for people. Figure 1 which is the statistic of the dataset reported by McAuley in [15] shows that the annual sales volume of the book domain under the Amazon shopping platform. It's obvious that the consuming volume are expending exponentially on the e-commerce platform. While bringing people diversified product choices, their growth also makes it more difficult for users to discover the products they are really interested in. How to help users discover the products they are interested in becomes a key topic in the development of an e-commerce platform. Efficiently helping user discover products of interest not only helps improve the user experience and increases user stickiness, but also stimulates consumption which could improve the profitability of platform. Recommender systems [1] which is considered as a kind of technology to solve the information overload problems, have been applied to product recommendation successfully on the past several decades.

© Springer Nature Switzerland AG 2018
S. Zhang et al. (Eds.): CCIR 2018, LNCS 11168, pp. 237–248, 2018.
https://doi.org/10.1007/978-3-030-01012-6_19

Current product recommendation technology mostly build user product rating matrix based on user rating information. Under such setting, the problem is modeled as matrix filling problems. And recommendation list is generated by sorted the predicted item of the rating matrix from high to low. Such technology has achieved certain success. However, with the development of e-commerce platforms, there are more and more ways for user to give feedback. Among such feedback, user could give review to the product to explain why they rated the product or they can express what feature the product has and what feature they like

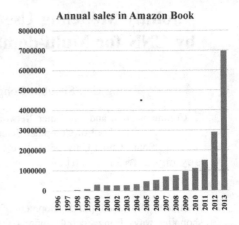

Fig. 1. The tendency of book sales in Amazon

and so many other useful information. Obviously, the recommender system could perform more efficiently by mining the user's preference and product's feature from the review text. On the other hand, the current recommendation technology mainly focuses on single domain, such as book recommendations, movie recommendations, music recommendations, separately. We argue that the user's consuming behavior between different domains has a certain degree of commonality which could enhance the effort of recommendation a lot, for example, if a user enjoys the "Harry Potter" movies, it's reasonable to infer that that user also likes Harry Potter's novels. By mining such commonality of user interests in multiple domains, recommender systems could generate a better recommendation for user. So, we needs to leverage the information mining from review text and the common consuming behavior from multi domain to achieve a better recommendation.

To touch this two points, this paper proposes a multi-domain recommendation algorithm based on CNN which is called CCoNN. It constructs the user preference vector and product feature vector by mining the common consuming behavior among different domains from the reviews text written by the same user. Based on these vectors, the user's rating of an unknown product is finally predicted.

The contributions of this paper are summaried as follow.

(1) A effective way to construct the representation vector of user preference and product characteristic is proposed, which leverage a CNN to transform the information mining from the user's review among multi-domain;

(2) The accuracy of product rating prediction is improved by leverage such representation vector based on Factorization machine for multi-domain recommendation.

The rest of the paper is organized as follows. The second part describes related work on recommender systems; the third part introduces the CCoNN in detail; the fourth part gives the result of the experiment and some analysis of it, followed by the conclusion and future work.

2   Related Works

The common approach of multi-domain recommendation [2] is to construct user preference representation by modeling user behavior in different domains, which benefits user cold start problem in single-domain recommendations. Vinayak [3] constructs the user's preference vector by leveraging the user's movie browsing record and the genre tag of the movie, and then books are recommended to the user whose favorite movie genre tag is similar. Shrivastva [4] constructs a tensors of users, items, and tags dimension by mining the tags information of users and recommendation are generated among multi-domain based on tensor decomposition. Li [5] treats the book and movie as different domains. They transform the rating pattern of user learned from movie domains into book domains and finally achieve better recommendation.

With the development of the e-commerce platform and the increasing of user review text, researchers have pay more attention on how to use these review text to improve the performance of recommendation in recent years. McAuley [6] proposed the HFT model by combining the LDA topic model and Matrix Factorization model, which corresponded the subject of the user review text with the latent factor vector of the rating. Their work has increased the interpretability of the recommendation by using the user review text. Ling [7] developed the RMR model to fit two assumptions, (1) user's rating behavior is the combination of the evaluation on product's different aspects, and (2) their evaluation on each aspect follows the Gaussian distribution whose parameter is relative to the review text. Finally, they use Gibbs sampling algorithm to acquire the arguments of the model. Compared with the HFT, RMR improved the precision of overall rating prediction without loss of interpretation. On the other hand, Deep Learning techniques has catched monay researchers' eyes. Many review based recommender systems using Deep Learning techniques have been proposed. Zheng et al. [8] proposed DeepCoNN to model user and product from reviews through two parallel CNN [9], finally a shared layer based on Factorization Machine [10] is used for rating prediction. However, their work mainly focuses on single-domain recommendation problems. This paper expands DeepCoNN to achieve multi-domain product recommendation which improves the accuracy of rating prediction under the multi-domain setting.

3   Rating Prediction Algorithm for Multi-domain Basic on CNN

In order to mining the user preference and product feature from user review text in multiple domains, this paper expands the DeepCoNN and proposes a multi-domain product rating prediction algorithm based on CNN, which is called CCoNN. The hold architecture of the algorithm is shown as Fig. 1. It consists of three independent components, text embedding layers and a convolutional neural network layer first, where the text embedding layer embeds every word into a unique vector, and every review was viewed as a sequence of words which was transformed into a sequence of vectors by the text embedding layer. The output vector of the text embedding layer are treated as the representation of the user or product. After that, we apply a convolutional neural network to capture the local information among a certain word range. And in the paper, we use several size of word range, specifically, we use different convolution kernel size to apply convolution operation on the same vector to capture the information more detailed. Following the convolution operation, a max pooling layer followed by a dense connected layer is applied to obtain the final output of the convolutional layer. Such output vectors is treated as user preference vector or a product feature vector in different domains. At the top of the architecture, similar with DeepCoNN, this paper uses two Factorization Machine to transform the concatenation of user preference vector and product feature vector into the rating prediction of the user and the product. Finally, the entire algorithm is trained by minimizing the loss between the predicted score and the true score.

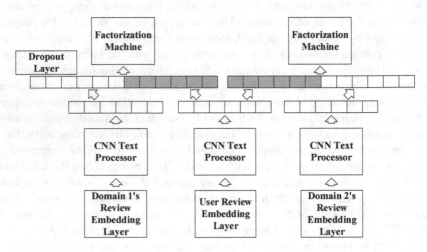

Fig. 2. The architecture of the algorithm

3.1 Review Embedding Layer

The user's review on the product is a kind of textual data which should be transformed into numerical value so that such review text could be the input of CNN. In recent years, the distributed representations of words [11, 12] has been successfully applied in NLP. This paper follow such idea to obtain the representation of word so that the review text belong to user or product could be represented as a sequence of such vector. For products, the review text of them in one domain could be sorted by time that they were written. Then the product could be viewed as a concactenation of such review text which is called document in this paper. As for users, since a user could write review for product under multi-domain, we sort the review written by he among every domain by the create time of the review to form his representation. Note that the commonality of a user would reflect in different domains through the review he writed, we could capture such information by such representation.

Formally, we define a mapping from a word to vector.

$$v = f(x), f : T \to R^n \tag{1}$$

Where T is the word set and R^n represents an n-dimensional vector space. Thus, each user can be represented in the order of words in his review document.

$$V_{m \times n}^u = \begin{bmatrix} f(x_1) \\ \vdots \\ f(x_m) \end{bmatrix}_{m \times n} \tag{2}$$

Among them, V is the output matrix of the text embedding layer whose shape is m × n. Here, m is the number of words in the current user document, and n is the length of the vectorization. Compared to the Bag-Word (BOW) approach, this representation can capture the time-order information of user reviews better and keep the order information of words.

3.2 CNN Text Processor

In order to capture the user preference and product feature information contained in the review text, an non-linear transformation is applied on $V_{m \times n}^u$ inside the CNN Text Processor layer. The CNN Text Processor layer is consis of a CNN convolutional layer, a max pooling layer and a full-connection layer sequencely. Finally, it forms the user preference vector or product feature vector. The internal structure of this layer is shown in Fig. 3, where the convolutional layers differ in the convolution kernel size it uses.

With the purpose of reserving the unique information of different words length, we setup different sizes of convolution kernel to apply the convolution operation on the same for $V_{m \times n}^u$ in the convolution layer. The convolution kernel of the i-th neuron was expressed as $K_i^{t \times n}, t \in F$, where F represents the set of kernel size, t is the convolutional size, and n indicates the size of word vector. With such setting, the output of the convolution operation could be represented as follow.

$$CO_t = \begin{bmatrix} z_{t1} & z_{t2} & \cdots & z_{tj} \end{bmatrix}_{(m-t+1) \times j} \tag{3}$$

$$z_{ti} = f\left(K_i^{t \times n} * V_{m \times n}^u + b_i\right), i \in [1, 2, \ldots, j] \tag{4}$$

Where t is the size of the current convolution kernel, m is the number of words in the user document, and j represents the number of output neurons set in the convolutional layer. Meanwhile, z_{ti} whose shape is $(m - t + 1) \times 1$ represents the output of the i-th neuron with the convolution kernel $K_i^{t \times n}$. And b_i is the global bias and f is an activation function to introduce a non-linear mapping relationship.

Here we uses the ReLU function as the f, which is defined as follows.

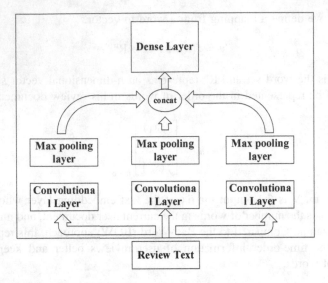

Fig. 3. The internal structure of CNN text processor

$$\text{ReLU}(x) = \max(0, x) \tag{5}$$

The output vector z_{ti} contains the user's preference information or the product's feature information. We think that the maximum element of the z_{ti} could capture the most important factor on which a user will consider when evaluate a product or a product's most important characteristic. Therefore, we uses a maximum pooling layer to handle z_{ti}.

$$o_{ti} = \max(z_{ti}), i \in [1, 2, \ldots, j] \tag{6}$$

$$O_{1 \times j}^t = [o_{t1}, o_{t2}, \ldots, o_{tj}], t \in F \tag{7}$$

Such operation not only retain the most important features, but also reduce the complexity of the calculation. Here, the $O_{1 \times j}^t$ is the output vector consis of the maximal element of every output neuron with the same convolution kernel size t. And for every t in F, all vector $O_{1 \times j}^t$ are concactened as $O_{1 \times s}$.

$$O_{1 \times s} = concatenate\left(O_{1 \times j}^{t_1}, O_{1 \times j}^{t_2}, \ldots, O_{1 \times j}^{t_i}\right), t_i \in F, s = sum(F) \tag{8}$$

Finally, the $O_{1 \times s}$ is passed through a dense layer which transforms it to the result of CNN Text Processor layer. The output vector x_{fui} is defined as the user preference vector or product feature vector in this paper.

$$x_{\text{fui}} = f(W \times O_{1 \times s} + b) \tag{9}$$

Among them, W is the weight matrix of the dense layer, b is the global bias, and f is the activation function. Similarly, we uses the ReLU function as the activation function here.

3.3   Multi-domain Rating Predictor Basic on FM

In recent years, Factorization Machine is more and more popular due to its superiority in modeling higher-order relationships between different features in various regression and classification problems. In this paper, the CNN Text Processor layer could be viewed as a function which maps user or product reviews into different implicit spaces which means that the output vector of user and product couldn't be used in the same spaces. In order to discovery the cross-relationship between the representative vector of user and product, Factorization Machine is used as the core of the last layer, the rating prediction layer, of the algorithm.

In our architecture, the review text from different domain was fed into the CNN Text Processor layer which generates user preference vector and product feature vector in different domain which was denoted as u_f, d_{1f}, d_{2f}, respectively. We argue that the user's preference in different domains should have commonality which benefit the rating prediction problem and it is shared in different domains, so this paper first connects user features with product features to construct the input vector for the Factorization Machine in different domain.

$$z_1 = concatenate\left(u_f, d_{1f}\right) \tag{10}$$

$$z_2 = concatenate\left(u_f, d_{2f}\right) \tag{11}$$

z_1, z_2 will be used as the input to the Factorization Machine in both domain. And the rating prediction p_i could be represented as (12).

$$p_i = b_0 + \sum_{j=1}^{|z_i|} w_i z_{ij} + \sum_{j=1}^{|z_i|} \sum_{k=j+1}^{|z_i|} <v_j, v_k> z_{ij} z_{ik}, i \in [1,2] \tag{12}$$

Among them, b_0 denotes the global deviation term, w_i denotes the weight of each element of the vector z_i, which is interpreted as the importance of each component of the user preference vector or the product feature vector on the final product rating, and v is used to model the cross-relationship between the elements of z_1 and z_2. $<v_j, v_k>$ represents an inner product operation. The prediction$_i$ is the score that the current user will rate for the product, where i specify the domain to which the current product belongs.

3.4　Objective Function

After we get the rating prediction, we hope that the different value between the prediction and the truth rating in the two domains will be as small as possible. Therefore, we define the following objective function.

$$\text{loss} = \sum_{i=1}^{N} (p_i - r_i)^2 \tag{13}$$

Where N is the size of train set, and r_i is the truth rating. This paper uses Adam [13] to minimize the loss. And we stipulate that the parameter of the network from a domain will be updated when only the train data is from that domain. In other words, if the current user and product reviews come from domain one, then only the parameters of product CNN Text Processor and the Factorization Machine of domain one will be updated. As for the user CNN Text Processor, it is shared by all domains, so the parameters of the it will be updated for every time.

4　Experiment and Analysis

The dataset used in this paper is the shopping dataset from Amazon, which was published by Julian McAuley [14, 15]. The dataset contains user consumption records for all domains in the Amazon e-commerce platform, such as books, movies, music, etc., with a time span of 1996.05–2014.07, which contains 142.8 million record in total. Each record is construct with the following field: reviewerID (user ID), asin (product ID), overall (user rating of the product), reviewText (user reviews on the product). The hole dataset is consists of 24 domain in Amazon.

4.1　Description of Dataset

The experiment in this paper mainly used six domains: "Automotive", "Musical_Instruments", "Beauty", "Digital_Music", "Toys_and_Games", "Clothing_ Shoes_ and_Jewelry". The simple statistics of them are shown in Table 1.

Table 1. Statistics of domain dataset

Domain	#user	#product	#record	Words every review
Musical_Instruments	1429	900	10261	91.67
Automotive	2928	1835	20473	86.96
Office_Products	4905	2420	53258	148.31
Tools_and_Home_Improvement	16638	10217	134476	111.87
Digital_Music	5541	3568	64706	199.91
Toys_and_Games	19412	11924	167597	101.92
Beauty	22363	12101	198502	90.08
Clothing_Shoes_and_Jewelry	39387	23033	278677	61.21

In this paper, we designed four groups of experiments which use two domains data as the same time. As shown in the above table, every two domains between the double-line frames are used in one experiment. For each experiment, we first turned the text of the reviews in the two domains into lowercase letters and changed the words that like he's, haven't, you're, I'll, into he's, haven't, you're., I' ll. After that, the vocabulary is constructed by splitting the space character. Every word of all domains is transform into a integer. With such mapping, the review text of word could be transformed into a sequence of integer. On the other hand, since the length of review text written by different user is different, we need to cut them into the same length. So, we first sorts the length of the user's review from small to large, takes its 85% quantile as the standard length. Review text which is larger than standard length will be truncated, and the others will be padding by a special character "<PAD>". Finally, the experimental data of this paper is generated, and the data sets of all domains are divided into training sets and test sets according to the ratio of 3:1.

4.2   Experiment Analysis

In this experiment, this paper selected the following algorithms as the baseline algorithm:

(1) DeepCoNN [8], a Deep Learning based algorithm which focuses on single domain recommendation. It mainly uses the CNN to transform the review text of user and product into vectors parallelly. Such vectors are concacted as the input vector of the rating prediction layer which use Factorization Machine to generate the rating prediction.

(2) NMF [16], a traditional single-domain rating prediction algorithm. The core of it is non-negative matrix factorization. It decomposes the user-product rating matrix into two submatrix. After getting such submatrix, a rating matrix is reconstructed by the inner product of them on which recommendation is generated based;

(3) KNNWithZScore [17], a kind of user-based collaborative filtering algorithm to score deviations, which calculates the cosine similarity of the user's rating vector among users in a single domain, and obtains a similar user list for the active user. The rating prediction of the active user for a certain product is generated based on the rating information of the list of similar users. In this experiment, the length of the similar user list is 40;

(4) KNN-CDCF [18], a multi-domain collaborative filtering algorithm based on similar users which treats the information from multi-domain as from single-domain. And it applys the traditional user-based CF algorithm to generate recommendation. In this paper, the length of the user's similar list is 40.

There are several parameters of CCoNN that need to be setup before the experiment, including the size of the convolution kernel in the convolutional network, the number of output neurons in the convolutional layer, and the length of vector v in (12). In this experiment, the settings of each parameter in this paper are shown as Table 2.

Table 2. The parameters of CCoNN

Parameters	Values
Convolution kernel size	[2, 3, 5]
Each convolution kernel size corresponds to the number of output neurons	[8, 8, 8]
The length of vector v in Factorization Machine	8

The above parameters are valid for all CNN Text Processor layers in the model. For fairness, we also use the same parameter settings for DeepCoNN. All models are trained on the same training set and evaluated on the same test set. Since the model proposed in this paper is mainly aimed at the problem of accuracy of scoring prediction, the evaluation index of the algorithm used in this experiment is RMSE. The specific calculation values are as follows.

$$RMSE = \frac{\sqrt{\sum_{i=1}^{len(testset)} (prediction_i - overall_i)^2}}{len(testset)} \tag{14}$$

The experimental results are shown in the following table.

Table 3. The RMSE of the algorithms

Domain	CCoNN	Deep CoNN	NMF	KNN With ZScore	KNN-CDCF
Musical_Instruments	**0.9190**	0.9692	1.079	0.996	0.9897
Automotive	**0.9597**	0.9663	1.087	1.014	1.0236
Office_Products	**0.9229**	0.9241	0.976	0.976	0.9280
Tools_Home_Improvement	**1.020**	1.028	1.153	1.095	1.0956
Digital_Music	1.006	1.004	1.038	**0.991**	1.0023
Toys_and_Games	0.9520	**0.9512**	1.054	1.007	1.0163
Beauty	**1.132**	1.1348	1.262	1.202	1.2059
Clothing_Shoes_and_Jewelry	1.072	**1.070**	1.244	1.153	1.1545

From the experimental results shown in Table 3, CCoNN is better than others in five of eight domains, while in the other domains, it is weaker than the best algorithm slightly. At the same time, compared to the traditional methods such as Matrix Factorization and Collaborative Filtering, the improvement of CCoNN is more obvious. It models user preference vectors and product feature vectors based on review text and rating information compared with such two traditional methods which calculated the implicit factor vector from the rating information only; at the same time, compared with DeepCoNN which could only be used in the single-domain rating prediction problem, we broken such limitation by extand its network structure through sharing user preference vectors in multiple domains which models the phenomenon that users' shopping preferences in different domains should have some kind of commonality. Such extension such benefit the accuracy of the rating prediction problem in multi-domain recommendation problem.

5 Conclusion

In the current e-commerce platform, review text, a kind of user feedback, will be more and more richer. How to leverage such review data to mine the user preference and product feature more accurately and deeply becomes a key challenge in product recommender system. In this paper, we proposed a CNN-based multi-domain rating prediction algorithm to address such problem. It first transform the review text in different domain into user preference vector and product feature vector by the CNN Text Processor. After that, by share the user preference vector among different domain, we fits our assumption that user's consumption behavior among multi-domain has a certain kind of commonality well. Then, such user preference vector are concatenated with the product feature vector to form the input vector of the rating prediction layer. Finally, we train such model by minimize the difference of the prediction rating and the truth rating under the setting of multi-domain. The experiments show that the algorithm has a better performance than the others.

The work described also has its shortcomings. The way to sharing user preferences among different domain should be explored in more depth. Some promising directions for further research include: Exploring advanced methods to share the user preference, such as modeling under preference in different domain separately and learning a mapping function among the models.

Acknowledgement. The research of this paper was supported by the Guangdong Science and Technology Key Project (2015B010131009), and Scientific Research Joint Funds of Ministry of Education of China and China Mobile (MCM20150512).

References

1. Resnick, P., Varian, H.R.: Recommender systems. Commun. ACM **40**(3), 56–58 (1997)
2. Cantador, I., Fernández-Tobías, I., Berkovsky, S., Cremonesi, P.: Cross-domain recommender systems. In: Ricci, F., Rokach, L., Shapira, B. (eds.) Recommender Systems Handbook, pp. 919–959. Springer, Heidelberg (2015). https://doi.org/10.1007/978-1-4899-7637-6_27
3. Vinayak, S., Sharma, R., Singh, R.: A personalized social network based cross domain recommender system. In: Corchado Rodriguez, J., Mitra, S., Thampi, S., El-Alfy, E.S. (eds.) ISTA 2016. AISC 2016, vol. 530, pp. 831–843. Springer, Heidelberg (2016). https://doi.org/10.1007/978-3-319-47952-1_66
4. Shrivastva, K.M.P., Singh, S.: Cross domain recommendation using semantic similarity and tensor decomposition. Procedia Comput. Sci. **85**, 317–324 (2016)
5. Li, B., Yang, Q., Xue, X.: Can movies and books collaborate? Cross-domain collaborative filtering for sparsity reduction. In: IJCAI, vol. 9, pp. 2052–2057, July 2009

6. McAuley, J., Leskovec, J.: Hidden factors and hidden topics: understanding rating dimensions with review text. In: Proceedings of the 7th ACM Conference on Recommender Systems, pp. 165–172. ACM, October 2013

7. Ling, G., Lyu, M.R., King, I.: Ratings meet reviews, a combined approach to recommend. In: Proceedings of the 8th ACM Conference on Recommender Systems, pp. 105–112. ACM, October 2014

8. Zheng, L., Noroozi, V., Yu, P.S.: Joint deep modeling of users and items using reviews for recommendation. In: Proceedings of the Tenth ACM International Conference on Web Search and Data Mining, pp. 425–434. ACM, February 2017

9. Collobert, R., Weston, J., Bottou, L., Karlen, M., Kavukcuoglu, K., Kuksa, P.: Natural language processing (almost) from scratch. J. Mach. Learn. Res. 12(Aug), 2493–2537 (2011)

10. Rendle, S.: Factorization machines. In: 2010 IEEE 10th International Conference on Data Mining (ICDM), pp. 995–1000. IEEE, December 2010

11. Bojanowski, P., Grave, E., Joulin, A., Mikolov, T.: Enriching word vectors with subword information (2016). arXiv preprint arXiv:1607.04606

12. Mikolov, T., Chen, K., Corrado, G., Dean, J.: Efficient estimation of word representations in vector space (2013). arXiv preprint arXiv:1301.3781

13. Kinga, D., Adam, J.B.: A method for stochastic optimization. In: International Conference on Learning Representations (ICLR), vol. 5 (2015)

14. He, R., McAuley, J.: Ups and downs: modeling the visual evolution of fashion trends with one-class collaborative filtering. In: Proceedings of the 25th International Conference on World Wide Web, pp. 507–517. International World Wide Web Conferences Steering Committee, April 2016

15. McAuley, J., Targett, C., Shi, Q., Van Den Hengel, A.: Image-based recommendations on styles and substitutes. In: Proceedings of the 38th International ACM SIGIR Conference on Research and Development in Information Retrieval, pp. 43–52. ACM, August 2015

16. Lee, D.D., Seung, H.S.: Algorithms for non-negative matrix factorization. In: Advances in Neural Information Processing Systems, pp. 556–562 (2001)

17. Su, X., Khoshgoftaar, T.M.: A survey of collaborative filtering techniques. Adv. Artif. Intell. 2009, 19 (2009)

18. Cantador, I., Cremonesi, P.: Cross-Domain Recommender Systems 2nd ed. of the RSs Handbook Cross-Domain Recommender Systems, October 2014

Sequence Modeling

Learning Target-Dependent Sentence Representations for Chinese Event Detection

Wenbo Zhang, Xiao Ding, and Ting Liu[✉]

Research Center for Social Computing and Information Retrieval,
Harbin Institute of Technology, Harbin, China
{wenbozhang,xding,tliu}@ir.hit.edu.cn

Abstract. Chinese event detection is a particularly challenging task in information extraction. Previous work mainly consider the sequential representation of sentences. However, long-range dependencies between words in the sentences may hurt the performance of these approaches. We believe that syntactic representations can provide an effective mechanism to directly link words to their informative context in the sentences. In this paper, we propose a novel event detection model based on dependency trees. In particular, we propose transforming dependency trees to target-dependent trees where leaf nodes are words and internal nodes are dependency relations, to distinguish the target words. Experimental results on the ACE 2005 corpus show that our approach significantly outperforms state-of-the-art baseline methods.

Keywords: Event detection · Neural network
Syntactic dependency tree

1 Introduction

Event detection (ED) refers to identifying event instances of specified types in plain texts. Each event mention is often presented within a single sentence in which an event trigger is selected to associate with that event mention. Event tiggers are generally single verbs or nominalizations that evoke the corresponding events. For instance, considering the sentence *S1* shown in Fig. 1, an ED system is expected to detect a *Be-Born* event along with the trigger word 出生 *(born)*. The event detection task, more precisely stated, involves identifying event triggers and classifying them into specific event types. This is an important and challenging task of information extraction in natural language processing (NLP), as the same event might appear in the form of various expressions and an expression might represent different events in different contexts.

There are two fundamental limitations for conventional feature-based methods of ED: *(i)* the complicated feature engineering and *(ii)* the error propagation from NLP toolkits and resources (i.e., word segmentation, part of speech tagging,

© Springer Nature Switzerland AG 2018
S. Zhang et al. (Eds.): CCIR 2018, LNCS 11168, pp. 251–262, 2018.
https://doi.org/10.1007/978-3-030-01012-6_20

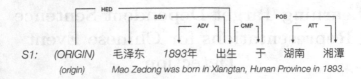

Fig. 1. Event mentions and syntactic analysis results for the example sentence.

parser etc.). To solve these issues, deep learning models, such as convolutional neural networks (CNN) (Chen et al. 2015; Nguyen and Grishman 2015, 2016a) and recurrent neural networks (RNN) (Nguyen et al. 2016b) are used for end-to-end ED. However, in the basic implementation, neural networks can only model consecutive context information in the sentences. Such consecutive mechanism is unable to capture the long-range and non-consecutive dependencies that are necessary to the prediction of triggers.

In the view of long-range dependencies, the non-consecutive CNN model (Nguyen and Grishman 2016a) seeks to solve this problem by operating the temporal convolution over all the non-consecutive k-grams in the sentences, achieving a breakthrough effect. However, the non-consecutive CNN models all possible non-consecutive k-grams that has high time complexity, and models unnecessary and noisy information, potentially impairing the prediction preformance. One way to circumvent this issue is to join syntactic dependency relationship and directly link words to their informative context in the sentences. This strategy not only overcomes long-range dependencies, but also alleviates the impacts of noisy information.

In order to make use of the syntactic structure information, we employ the tree-structured long short-term memory networks (Tree-LSTM) (Tai et al. 2015; Chang et al. 2016), which use tree structures to form connections between layers of neural networks. In Tree-LSTM, the node vector is computed from the representation vectors of its subnodes. Tree-LSTM has been mainly applied for the classification tasks in which the hidden representation vector of the root node is used as the sentence features for classification, just meeting our task requirements for ED. Meanwhile, we employ bidirectional LSTM networks whose hidden outputs are used as the inputs of Tree-LSTM to model sentence-level information.

Our main contributions are three folds: *(1)* We propose a novel neural network model for Chinese event detection task, which tackles long-range dependencies by joining the syntactic dependency relationship. To our konwledge, this is the first neural network model over dependency tree for Chinese event detection task. *(2)* We systematically investigate the effect of different network architectures for ED. *(3)* We conduct extensive experiments on the ACE 2005 corpus, and the experimental results show that our approach achieves the best performances comparing with state-of-the-art baseline methods.

2   Task Description

The event detection task we investigating is that of the Automatic Content Extraction (ACE) evaluations LDC (2005), where an event is defined as a specific occurrence involving one or more participants. Event detection task requires that certain specified types of events should be detected. We first introduce some ACE terminologies to facilitate the understanding of the task:

- Entity: an object or a set of objects in one of the semantic categories of interest.
- Entity mention: a reference to an entity (typically, a noun phrase).
- Event mention: a phrase or sentence within which an event is described, including trigger and arguments.
- Event trigger: the main word which most clearly expresses an event occurrence.
- Event arguments: the entity mentions that are involved in an event.
- Event type: a particular event category.

The ACE 2005 evaluation has 8 types of events, with 33 subtypes. In this paper, we will treat these simply as 33 separate event types and ignore the hierarchical structure. Besides, instead of doing the overall event extraction task, we concentrate only on the event detection task – namely Event Trigger Identification and Event Type Classification.

3   Methodology

In this section, we illustrate the details of our approach. Similar to existing work, we formalize ED as a multi-class classification task. Given a sentence, for each token in the sentence, we want to predict whether the current token is an event trigger in the pre-defined event set or not? Specifically, our goal is to classify each word in the sentence into one of 34 classes (33 event types plus an *NONE* class).

Our proposed neural network architecture is shown in Fig. 2, consisting of three modules: *(i)* the encoding module that represents the input sentence with a matrix for neural network computation, *(ii)* the bidirectional LSTM module that models sentence-level information, *(iii)* the tree-structured LSTM module that models syntax-level information to perform ED.

3.1   Encoding

In the encoding module, each token w_i in the input sentence $S = w_1, w_2, ..., w_n$ is transformed into a real-valued vector x_i by concatenating the following vectors:

- The word embedding vector of w_i: Word embeddings are able to capture the meaningful semantic regularities of words, which are often pre-trained on some large unlabeled corpora. We obtain word embeddings by looking up a pre-trained word embedding table.

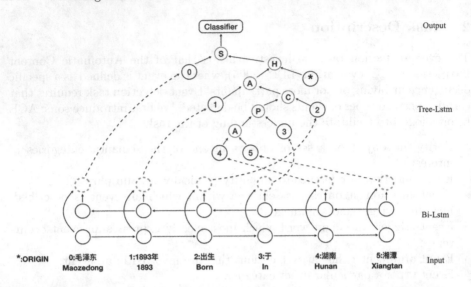

Fig. 2. The architecture of our proposed neural network model in which leaf nodes and internal nodes of Tree-LSTM are words and dependency relations, respectively. In particular, *ORIGIN* is randomly initialized and optimized through back propagation.

- The position embedding vector of w_i: In order to indicate which word is the current word, we encode the relative distance from context words to the current word as a real-valued vector (called as the position embedding vector) and use this vector as an additional representation of w_i. For example, in *S1*, the relative distances of 毛泽东*(Maozedong)* to the trigger 出生*(born)* is 2. To encode the position feature, each distance value is also represented by an embedding vector, which is randomly initialized and optimized through back propagation.

   As each token w_i is represented by the vector x_i, the input sentence S can be seen as a sequence of vectors $X = x_1, x_2, ..., x_n$. X would be used as the input for the bidirectional LSTM networws in the next step.

3.2   Bidirectional LSTM

In this paper, to abstract the initial representation, we employ a bidirectional long-short term memory network (BiLSTM) (Hochreiter and Schmidhuber 1997) to obtain hidden vectors whose outputs are later consumed by Tree-Structured LSTM.

   Specifically, we run a forward LSTM and a backward LSTM over the representation vector sequence $(x_1, x_2, ..., x_n)$ to generate the forward and backward hidden vector sequences (i.e., $(\overrightarrow{h_1}, \overrightarrow{h_2}, ..., \overrightarrow{h_n})$ and $(\overleftarrow{h_1}, \overleftarrow{h_2}, ..., \overleftarrow{h_n})$) respectively). We then concatenate the hidden vectors at the corresponding positions to obtain the abstract representation vector sequence $(h_1, h_2, ..., h_n)$ where $h_i = [\overrightarrow{h_i}, \overleftarrow{h_i}]$, which would replace the initial vector sequences $(x_1, x_2, ..., x_n)$ for the further computation of Tree-LSTM.

3.3   Tree-Structured LSTM

A limitation of the LSTM architecture is that they only allow for strictly sequential information propagation. However, natural language exhibits syntactic properties that would naturally combine words to phrases. Therefore, we use the Tree-LSTM (Tai et al. 2015), which generalizes LSTM to tree-structured network topologies and preserves syntactic information.

Just like a standard LSTM unit, each Tree-LSTM unit also contains input and output gates i_j and o_j, a memory cell c_j and hidden state h_j, but the difference between them is that gating vectors and memory cell updates of the Tree-LSTM unit are dependent on the states of possibly many child units. Moreover, instead of a single forget gate, the Tree-LSTM unit contains one forget gate f_{jk} for each child k. This allows the Tree-LSTM unit to selectively incorporate information from each child.

Given a dependency tree, let $C(j)$ denotes the set of children of node j. The Tree-LSTM transition equations are the following:

$$\tilde{h}_j = \sum_{k \in C(j)} h_k \tag{1}$$

$$i_j = \sigma(W^{(i)}x_j + U^{(i)}\tilde{h}_j + b^{(i)}) \tag{2}$$

$$f_{jk} = \sigma(W^{(f)}x_j + U^{(f)}h_k + b^{(f)}) \tag{3}$$

$$o_j = \sigma(W^{(o)}x_j + U^{(o)}\tilde{h}_j + b^{(o)}) \tag{4}$$

$$u_j = \tanh(W^{(u)}x_j + U^{(u)}\tilde{h}_j + b^{(u)}) \tag{5}$$

$$c_j = i_j \odot u_j + \sum_{k \in C(j)} f_{jk} \odot c_k \tag{6}$$

$$h_j = o_j \odot \tanh(c_j) \tag{7}$$

where $k \in C(j)$. Intuitively, we can interpret each parameter matrix in these equations as encoding correlations between the component vectors of the Tree-LSTM unit, the input x_j, and the hidden states h_k and the cell states c_k of the unit's children.

Tree-LSTM structure mentioned above allows for richer network topologies where each LSTM unit is able to incorporate information from multiple child units. However, the target-dependent tree that would be introduced following is a bifurcated structure, so we use the binary-tree-structured LSTM networks in this paper.

3.4   Target-Dependent Tree

For a particular sentence *S1* shown in Fig. 1, its syntactic dependency tree is unique as shown in Fig. 3a. The root is *ORIGIN* and head words are at the upper ends of dependency branches. In other words, the Tree-LSTM model would output only one representation for each word in the sentence, which is obviously irrational. Intuitively, there should be different dependency tree structures for different target words. Guided by this intuition, we propose a recursive algorithm (Algorithm 1) to transform a syntactic dependency tree T_s into a target-dependent tree T_t, given a target node n. The target-dependent tree obtained by recursive algorithm can highlight the importance of target words and better represent contextual information.

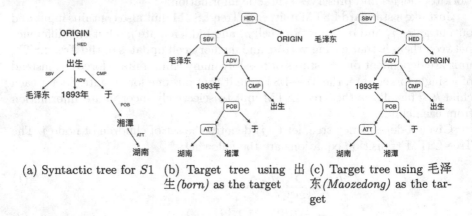

(a) Syntactic tree for $S1$ (b) Target tree using 出 生 *(born)* as the target (c) Target tree using 毛泽 东 *(Maozedong)* as the target

Fig. 3. Syntatic dependency tree and target-dependent tree

The algorithm starts from a given target node n. If the target node has a parent node, we create a bifurcated node to represent a dependency relation between the target node and its parent node, with the target node and parent node placed at the left and right children, respectively. If the target node has children, we create a bifurcated node to represent a dependency relation between the target node and its child node, with the child node and target node placed at the left and right children, respectively. After finishing a bifurcated node, we remove used nodes and the children nodes become new target nodes to expand the binary tree recursively until all nodes are transformed. In particular, the rule of selecting nodes is that the parent node is preceded by the child nodes, and the left context nodes are in front of the right context nodes. After transformation, leaf nodes of the target-dependent tree are words, and the internal nodes represent dependency relations as shown in Fig. 3b.

The target-dependent tree obtained by the above algorithm is a bifurcated structure, so we employ a binary Tree-LSTM model including two forget gates (f_{jl}, f_{jr}) in each Tree-LSTM unit. Given a dependency tree, let the node kl and kr denote the left and right children of node j, respectively. The left child unit states (h_{kl}, c_{kl}) corresponds to the left forget gate f_{jl}, and the right child unit states (h_{kr}, c_{kr}) corresponds to the right forget gate f_{jr}.

For a binary Tree-LSTM model running on a target-dependent tree, information is propagated from the bottommost leaf nodes to the topmost dependency node, and the final output is treated as the target-dependent sentence embedding for triggers identification and classification.

Algorithm 1. Syntactic-tree to Target-tree

Input: syntactic dependency tree T_s, target node n
Output: target-dependent tree T_t
1:   $T_t \leftarrow Transform(T_s, n)$
2:   $Transform(T_s, n)$ {
3:   **if** $T_s.HasParent(n)$ **then**
4:      $p \leftarrow T_s.GetParent(n)$
5:      $T_s.RemoveNode(p, n)$
6:      $tn \leftarrow NewTreeNode$
7:      $tn.LChild \leftarrow Transform(T_s, n)$
8:      $tn.RChild \leftarrow Transform(T_s, p)$
9:      **return** tn
10: **else if** $T_s.HasChild(n)$ **then**
11:      $c \leftarrow T_s.GetOneChild(n)$
12:      $T_s.RemoveNode(c, n)$
13:      $tn \leftarrow NewTreeNode$
14:      $tn.LChild \leftarrow Transform(T_s, c)$
15:      $tn.RChild \leftarrow Transform(T_s, n)$
16:      **return** tn
17: **else**
18:      **return** n
19: **end if**
20: }

3.5   Training

We train the network by minimizing the loss function values, and the optimization objective function is defined as the multi-class cross-entropy loss:

$$J(\theta) = - \sum_{i=1}^{N} \log p(y_i | t_i, \theta) + \lambda(\theta) \qquad (8)$$

where N indicates the number of all tokens in the training data. y_i is the true event type of token t_i. λ is the regularization parameter and θ indicates all parameters.

For optimization, we adopt adaptive moment estimation (Adam) (Kingma and Ba 2015) update rule to minimize the objective function. The gradients are computed via back-propagation while dropout is employed to avoid overfitting (Srivastava et al. 2014). Negative sampling is adopted to tackle with data imbalance problem. During training, besides the weight matrices, we also optimize the position embedding table to achieve the optimal states.

4   Experiments

4.1   Datasets and Evaluation Metrics

We utilize the standard ACE 2005 corpus for our experiments, which contains 633 Chinese documents. For the purpose of comparison, we use the same data split as previous work (Feng et al. 2016). This data split includes 60 documents for the test set, 60 other documents for the development set and 513 remaining documents for the training set.

We follow the criteria of previous work for evaluation: (1) A trigger is correctly identified if its offset matches a reference trigger (Event Trigger Identification). (2) A trigger is correctly classified if its trigger type and offsets exactly match a reference trigger (Event Type Classification). Precision (P), recall (R), and F1 score (F1) are used as the evaluation metrics, same as previous work for meaningful comparison.

4.2   Parameters and Resources

Weight matrix parameters are randomly initialized with uniform samples from [−0.1, 0.1]. Bias vectors are initialized to zero. Hyper-parameters are tuned on the development data of the ACE 2005 dataset. We employ pre-trained word embeddings with 200 dimensions from (Mikolov et al. 2013) to initialize the word embeddings. In addition, we set the dimension of the position embeddings to 50, the size of hidden layer to 100, the size of output layer to 34, the dropout rate to 0.5. These hyper-parameters are used in all experiments in this paper.

In order to parse the sentences in the datasets, we employ Language Technology Platform (Liu et al. 2011) for word segmentation, part of speech tagging and parsing on Chinese documents.

4.3   Baselines

In order to show the effectiveness, we compare between the proposed model (called B-TreeLSTM) and the state-of-the-art Chinese event detection systems on the ACE 2005 corpus. These baseline systems include:

- *Word-MEMM* (Chen and Ji 2009): the segmentation-based method that applies Maximum-Entropy Markov Model for ED.
- *MaxEnt* (Li et al. 2013): the pipeline model that employs human-designed lexical and syntactic features.
- *BUEES* (Ding et al. 2013b): a completely unsupervised way for event extraction and automatically building event type paradigm by clustering event triggers.
- *Char-MEMM* (Chen and Ji 2009): the first character-based method to handle the language specific issue, which trains a Maximum-Entropy Markov Model to label each character with BIO tagging scheme.
- *Char-based C-BiLSTM* (Zeng et al. 2016): a character-based sequence labeling model including bidirectional LSTM and CNN.
- *Word-based C-BiLSTM* (Zeng et al. 2016): a segmentation-based convolution bidirectional LSTM model that can capture both sentence-level and lexical features from raw texts.

- *HNN* (Feng et al. 2016): a hybrid neural network model that incorporates both bidirectional LSTMs and convolutional neural networks to capture sequence and structure semantic information from specific contexts for event detection.
- *Rich-L* (Chen and Ng 2012): a joint-learning and knowledge-rich approach that also incorporates Chinese-specific features, which is the feature-based state-of-the-art system.

In particular, for a fair comparison, we chose the pure deep learning model *C-BiLSTM* without errata tables. *Rich-L* is the best feature engineering method, which performs 10-fold cross-validation experiments.

4.4 Experimental Results

Table 1 shows the results of the comparison between the proposed model and baselines. From the table, we find that B-TreeLSTM significantly outperforms all the compared models. As a single model, it still outperforms the hybrid model (Feng et al. 2016). In addition, compared with the feature-based approaches, all the neural network models achieve prominent results. This is mainly because deep neural network can capture semantic and syntactic information in the absence of feature engineering and avoid the errors propagated from other NLP tasks. Tree-structured LSTM networks are superior to the sequential representation of models, as it can capture the long-range and non-consecutive dependencies over dependency trees.

Table 1. Comparison with existing Chinese event detection methods (%)

Model	Trigger identification			Trigger classification		
	P	R	F1	P	R	F1
Word-MEMM	68.1	52.7	59.4	65.7	50.9	57.4
MaxEnt	50.0	**77.0**	60.6	47.5	**73.1**	57.6
BUEES	n/a	n/a	n/a	72.7	50.7	59.7
Char-MEMM	**82.4**	50.6	62.7	**78.8**	48.3	59.9
Char-based C-BiLSTM	65.5	66.7	66.1	60.0	60.9	60.4
Word-based C-BiLSTM	75.8	59.0	66.4	69.8	54.2	61.0
HNN	74.2	63.1	68.2	77.1	53.1	63.0
Rich-L	62.2	71.9	66.7	58.9	68.1	63.2
B-TreeLSTM	68.7	73.2	**70.9**	62.1	65.8	**63.9**

In addition to B-TreeLSTM, we evaluate different LSTM architectures (i.e., BiLSTM and TreeLSTM) for ED. For TreeLSTM, we directly use the initial vector sequences $(x_1, x_2, ..., x_n)$ as the input. Moreover, there is a difference that BiLSTM is sequence labeling model, and the other two LSTM architectures are multi-class classification models.

Table 2 shows the performances of different LSTM architectures. The table indicates that B-TreeLSTM achieves the best performance compared with the other two models. It is precisely because B-TreeLSTM can simultaneously model sentence-level and syntax-level information, which tackles long-range dependencies. The comparison results also demonstrate the effectiveness of Tree-structured LSTM for ED.

Table 2. Comparison of different network architectures (%)

Model	Trigger identification			Trigger classification		
	P	R	F1	P	R	F1
BiLSTM	65	57.2	60.9	59.4	51.3	55.1
TreeLSTM	**69.1**	63.8	66.3	**64.2**	57.7	60.8
B-TreeLSTM	68.7	**73.2**	**70.9**	62.1	**65.8**	**63.9**

5   Related Work

Event detection has attracted great research efforts in recent years. The early approach for ED has involved the feature-based methods that rely on discriminative features to build statistical models. (Ahn 2006) leverages lexical features and syntactic features. To capture more clues from the texts, many advanced features were proposed, including cross-document features (Ji and Grishman 2008), cross-lingual features (Ji 2009), cross-event features (Liao and Grishman 2010), cross-entity features(Hong et al. 2011). (Li et al. 2013) present a joint framework to extract event triggers and event arguments with global features simultaneously. With the success of the deep learning, representation-based approaches have been introduced into ED recently. The typical models employs CNNs (Chen et al. 2015; Nguyen and Grishman 2015, 2016a), RNNs (Nguyen et al. 2016b) and attention-based networks (Liu et al. 2017, 2018). However, none of these works consides syntax for ED, until (Nguyen and Grishman 2018) put forward Graph Convolutional Networks that join syntactic structure information for the first time. Representation-based methods achieve relatively high performance because they can automaticly capture features and model complicated hidden interactions in data.

Compared with the rich research of event detection in English, the research of Chinese event detection receives less attention. (Chen and Ji 2009) first employ character-based method to handle the specific issue in Chinese. (Chen and Ng 2012) investigate a joint-learning and knowledge-rich approach to improve Chinese event detection. (Ding et al. 2013a) tackle the problem of domain adaptation in event extraction by automatically building event type paradigms based on the trigger clusters. (Ding et al. 2013b) present a bottom to up event extraction system, wich extracts events from the web in a completely unsupervised way. (Feng et al. 2016) propose a hybrid neural network model, which incorporates both bidirectional LSTMs and CNNs to capture sequence and structure semantic information from specific contexts, for Chinese event detection. (Zeng et al. 2016) introduce a convolution bidirectional LSTM model on Chinese event detection task. However, this paper is the first to integrate syntactic information in the neural network models for Chinese event detection.

6 Conclusion

In this paper, we propose a novel Chinese event detection method, which can tackle long-range dependencies based on tree-structured long short-term memory networks over dependency trees. Our approach is the first to integrate syntactic information in the deep learning models for Chinese event detection, which can capture sentence-level and syntactic-level features from raw texts. In addition, we propose a recursive algorithm to transform a syntactic dependency tree into a target-dependent tree given a target node. Experimental results show that our approach achieves significant performances on the ACE 2005 corpus.

References

Tai, K.S., Socher, R., Manning, C.D.: Improved semantic representations from tree-structured long short-term memory networks. In: Proceedings of ACL, pp. 1556–1566 (2015)

Chen, Z., Ji, H.: Language specific issue and feature exploration in Chinese event extraction. In: Proceedings of NAACL, pp. 209–212 (2009)

Chen, C., Ng, V.: Joint modeling for Chinese event extraction with rich linguistic features. In: Proceedings of COLING (2012)

Ding, X., Qin, B., Liu, T.: Building Chinese event type paradigm based on trigger clustering. In: Proceedings of IJCNLP, pp. 311–319 (2013a)

Ding, X., Qin, B., Liu, T.: BUEES: a bottom-up event extraction system. Front. Inf. Technol. Electron. Eng. 16, 541–552 (2013b)

Ding, X., Zhang, Y., Liu, T., Duan, J.: Using structured events to predict stock price movement: an empirical investigation. In: Proceedings of EMNLP, pp. 1415–1425 (2014)

Ding, X., Zhang, Y., Liu, T., Duan, J.: Deep learning for event-driven stock prediction. In: Proceedings of IJCAI (2015)

Chang, C.Y., Zhang, Y., Teng, Z., Bozanic, Z., Ke, B.: Measuring the information content of financial news. In: Proceedings of COLING, pp. 3216–3225 (2016)

Feng, X., Huang, L., Tang, D., Qin, B., Ji, H., Liu, T.: A language-independent neural network for event detection. In: Proceedings of ACL, pp. 66–71 (2016)

Zeng, Y., Yang, H., Feng, Y., Wang, Z., Zhao, D.: A convolution BiLSTM neural network model for Chinese event extraction. In: Proceedings of NLPCC (2016)

Liu, T., Che, W., Zhenghua, L.: Language technology platform. J. Chin. Inf. Process. 25(6), 53–62 (2011)

Ahn, D.: The stages of event extraction. In: Proceedings of ARTE, pp. 1–8 (2006)

Ji, H., Grishman, R.: Refining event extraction through unsupervied cross-document inference. In: Proceedings of ACL, pp. 254–262 (2008)

Ji, H.: Cross-lingual predicate cluster acquisition to improve bilingual event extraction by inductive learning. In: Proceedings of the NAACL HLT Workshop on Unsupervised and Minimally Supervised Learning of Lexical Semantics, pp. 27–35 (2009)

Liao, S., Grishman, R.: Using document level cross-event inference to improve event extraction. In: Proceedings of ACL, pp. 789–797 (2010)

Hong, Y., Zhang, J., Ma, B., Yao, J., Zhou, G., Zhu, Q.: Using cross-entity inference to improve event extraction. In: Proceedings of ACL (2011)

Li, Q., Ji, H., Huang, L.: Joint event extraction via structured prediction with global features. In: Proceedings of ACL, pp. 73–82 (2013)

Chen, Y., Xu, L., Liu, K., Zeng, D., Zhao, J.: Event extraction via dynamic multi-pooling convolutional neural networks. In: Proceedings of ACL, pp. 167–176 (2015)

Nguyen, T.H., Grishman, R.: Event detection and domain adaptation with convolutional neural networks. In: Proceedings of ACL, pp. 365–371 (2015)

Nguyen, T.H., Grishman, R.: Modeling skip-grams for event detection with convolutional neural networks. In: Proceedings of EMNLP, pp. 886–891 (2016a)

Nguyen, T.H., Cho, K., Grishman, R.: Joint event extraction via recurrent neural networks. In: Proceedings of NAACL, pp. 300–309 (2016b)

Liu, S., Chen, Y., Liu, K., Zhao, J.: Exploiting argument information to improve event detection via supervised attention mechanisms. In: Proceedings of ACL, pp. 2134–2143 (2017)

Nguyen, T.H., Grishman, R.: Graph convolutional networks with argument-aware pooling for event detection. In: Proceedings of AAAI (2018)

Liu, J., Chen, Y., Liu, K., Zhao, J.: Event detection via gated multilingual attention mechanism. In: Proceedings of AAAI (2018)

Hochreiter, S., Schmidhuber, J.: Long short-term memory. Neural Comput. 9(8), 1735–1780 (1997)

Mikolov, T., Sutskever, I., Chen, K., Corrado, G., Dean, J.: Distributed representations of words and phrases and their compositionality. In: Proceedings of NIPS, pp. 3111–3119 (2013)

Srivastava, N., Hinton, G., Krizhevsky, A., Sutskever, I., Salakhutdinov, R.: Dropout: a simple way to prevent neural networks from overfitting. J. Mach. Learn. Res. 15, 1929–1958 (2014)

Kingma, D.P., Ba, J.L.: ADAM: a method for stochastic optimization. In: Proceedings of ICLR (2015)

Prior Knowledge Integrated
with Self-attention for Event Detection

Yan Li[1], Chenliang Li[1], Weiran Xu[1(✉)], and Junliang Li[2]

[1] Beijing University of Posts and Telecommunications, Beijing, China
{liyan_12_xxgch,chenliangli,xuweiran}@bupt.edu.cn
[2] Luoyang Electronic Equipment Test Center, Luoyang, China
344036917@qq.com

Abstract. Recently, end-to-end models based on recurrent neural networks (RNN) have gained great success in event detection. However these methods cannot deal with long-distance dependency and internal structure information well. They are also hard to be controlled in process of learning since lacking of prior knowledge integration. In this paper, we present an effective framework for event detection which aims to address these problems. Our model based on self-attention can ignore the distance between any two words to obtain their relationship and leverage internal event argument information to improve event detection. In order to control the process of learning, we first collect keywords from corpus and then use a prior knowledge integration network to encode keywords to a prior knowledge representation. Experimental results demonstrate that our model has significant improvement of 3.9 F1 over the previous state-of-the-art on ACE 2005 dataset.

Keywords: Event detection · Self-attention
Prior knowledge integration

1 Introduction

Event detection is a crucial part of event extraction which also includes event argument extraction. We address the task of event detection: identifying event triggers and categorizing them. Event triggers with specific types evoke corresponding events. For example, consider the following sentence in the ACE 2005 dataset: *He died in the hospital*. An event detection system is expected to identify an *Die* event triggered by word *died*.

This task is quite difficult since the same event might be triggered by various trigger expressions and a trigger word might represent different events in different sentences. These problems lead to event detection task is closely related to syntax. Therefore traditional methods for event detection add many external syntactic features, which increases the complexity of system and brings extra error from NLP toolkits [7]. Recently, strategies without syntax inputs based on deep learning are widely used. These methods involve that end-to-end models with RNN can achieve state-of-the-art results in event detection task [9].

© Springer Nature Switzerland AG 2018
S. Zhang et al. (Eds.): CCIR 2018, LNCS 11168, pp. 263–273, 2018.
https://doi.org/10.1007/978-3-030-01012-6_21

RNNs show great potential in sentence sequence task. These approaches treat a sentence as a sequence and recursively integrate all words information before the current word. This feature makes RNNs to solve sequence-prediction tasks of any length theoretically. However, there still are many problems in practice. The main problem is the long-distance dependency [3]. RNNs cannot capture distant information, which leads to a decrease in the effect of sequence prediction. In addition, RNNs are less likely to learn the internal structure of sentence. Therefore, RNNs lack a way to tackle the task which requires a high level syntactic structure information.

The end-to-end model has achieved remarkable results in the event detection task and reveals the potential ability of capturing the contextual information [9]. Despite these success, without prior knowledge, end-to-end model just get the source sentence as input and then output the detection result, which certainly leads to the fact that the learning process cannot be controlled.

In order to effectively solve the above problems, we propose a effective model based on self-attention mechanism and prior knowledge integration for event detection. Our model use self-attention mechanism to learn sentence internal structure information. In particular, the relationship between arguments and triggers can be learned, so that the event argument information can be fully utilized in the event detection task. The most significant advantage of self-attention is that it can ignore the distance between any two words in a sentence. Therefore the long-distance dependency of RNNs can be solved. Besides, our model comes with a variant of RNN to further enhance the representations. In order to control the process of learning, we first collect a set of keywords from corpus and these words often appear as triggers in sentences. Then use a prior knowledge integration network to encode keywords to a prior knowledge representation. Our model applies the external representation to reduce the interference of redundant information and improves the event detection. Although the model is fairly simple, it gives remarkable empirical results.

2   Related Work

Event Detection. Event detection is a challenging task of event extraction. Existing methods mainly be divided into two groups. The first kind of approach tackled this task independently and totally ignored argument information. Early methods are feature-based, which exploited a lot of strategies to convert semantic features into feature vectors [1,5,8]. Recently, most of methods are based on sentence-level representations. In particular, [10] tackled event detection via CNNs and [2] apply dynamic multi-pooling CNNs to build model.

The second kind of approach tackled event detection and argument extraction simultaneously. [7,14,15] proposed structured perceptron. [15–17] used dual decomposition to solve event extraction. [9] applied RNNs to extract event jointly and achieved state-of-the-art result. However, none of these works utilizes self-attention and prior knowledge integration to perform event detection as we do in this work.

Self-attention. Self-Attention have been successfully used in several tasks. [3] used self-attention to solve the task of machine reading. [11] used self-attention to complete the task of natural language inference. [12] based on self-attention and reinforcement learning to achieve abstractive summarization. [19] applied self-attention to neural machine translation and obtained the state-of-the-art results. Recently, [18] applied self-attention to semantic role labeling task and achieved the state-of-the-art results. Our model apply self-attention to learn the long distance dependency and the internal structure information of sentence. Our experiments show the effectiveness of self-attention mechanism on event detection task.

3   Method

Sequence labeling methods have many applications in information extraction tasks [18, 20]. We formalize the event detection problem as a sequence labeling problem based on self-attention mechanism and prior knowledge integration. In this section, we firstly introduce how to change the event detection problem to a sequence labeling problem based on our model. Then we detail the model used to complete the task.

3.1   The Labeling Scheme

Figure 1 is an example of how the results are labeled. Each word in the sentence is assigned a label that contributes to extract the results. "O" label represents its corresponding token is not a trigger and other labels are event types. Event types information are already defined in the dataset. Therefore, the total number of labels is $N = |A| + 1$, where $|A|$ indicates the total number of event types.

Figure 1 shows the result of a sequence label extraction. The output is: {Trigger word: appointments, Event type:Contact.Meet}, where "Contact.Meet" represents the type of the trigger word "appointments".

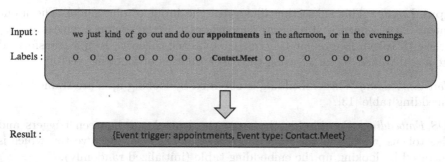

Fig. 1. An example sentence based on our labeling scheme.

3.2 Model

In the subsection, we introduce the model in detail. Our model is composed of N identical layers as Fig. 2 shows. Each layer contains a non-linear sublayer and a self-attention sublayer. We build a residual connection between each two sublayers, followed by layer normalization [6]. The final output of each layer is $LayerNorm(x + Sublayer(x))$. To help control the process of learning effectively, a prior knowledge integration network is added in the model's first layer. Finally, we take the outputs of the topmost attention sublayer as inputs to make the final predictions.

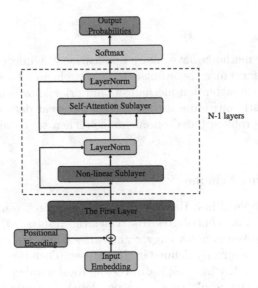

Fig. 2. An illustration of our framework based on self-attention mechanism and prior knowledge integration. The sentence embedding matrix is model's input.

Input Layer. This subsection illustrates the input of our model. The input sentence is a sequence $W = w_1 w_2 ... w_n$, where n is the sentence length and w_i is the i-th token. Before entering our model, each token w_i is first transformed into a real-valued vector x_i using the concatenation of following vectors:

Word Embedding: This embedding is obtained by looking up a pre-trained word embedding table [13].

POS Embedding: Considering the semantic dependency between triggers and other tokens, we encode the POS information of each word to a vector, which is generated by looking up the embedding table (initialized randomly).

The transformation from the token w_i to the vector x_i essentially converts the input sentence W into a sequence of real-valued vectors $X = (x_1, x_2, ..., x_n)$, to be used by our model.

Self-attention Sublayer. Self-attention is an attention mechanism that can learn sentence structure information and get sentence representation without external information. Self-attention has been used successfully in many tasks including reading comprehension, abstractive summarization, textual entailment, translation and semantic role labeling. To the best of our knowledge, our model is the first model relying on self-attention to solve event detection task [3,11,12,18,19].

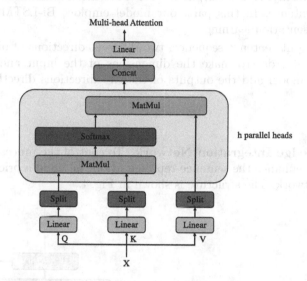

Fig. 3. The computation graph of multi-head self-attention mechanism.

In this paper, we mainly use multi-head attention mechanism to build model [19] . The structure of the mechanism is shown in the Fig. 3. The input dimension of this mechanism is d_{model}. This mechanism first linearly projects the queries, keys and values h times with different linear projections, respectively. Then h parallel heads are employed to focus on different representation subspaces at different positions.

The core of the multi-head attention mechanism is the module to compute the attention, which is called Scaled Dot-Product Attention. The input consists of queries and keys of dimension d_k, and values of dimension d_v. In practice, given a matrix Q, K, V. We compute the output of Scaled dot-product attention based on the following equation:

$$Attention(Q, K, V) = softmax\left(\frac{QK^T}{\sqrt{d}}\right)V \qquad (1)$$

Finally, all the outputs computed by h parallel heads are concatenated together. The information from different subspaces are merged by a linear projection. And the output of linear projection serve as the final output of the multi-head attention mechanism.

$$MultiHead(Q, K, V) = Concat(head_1, ..., head_h) \qquad (2)$$
$$where\ head_i = Attention(QW_i^Q, KW_i^K, VW_i^V) \qquad (3)$$

where the projections are parameter matrices $W_i^Q \in \mathbb{R}^{d_{model} \times d_k}$, $W_i^K \in \mathbb{R}^{d_{model} \times d_k}$ and $W_i^V \in \mathbb{R}^{d_{model} \times d_v}$.

Non-linear Sublayer. The output vectors of attention mechanism cannot fully express the sentence. In this part, our model employs Bi-LSTM to enhance sentence representation learning.

Given an input sentence sequence, two different directions of output $\overrightarrow{h_t}$, $\overleftarrow{h_t}$ are obtained. In order to make the dimensions of the input and output are the same, our model add the outputs of the two directions directly instead of concatenation:

$$y_t = \overrightarrow{h_t} + \overleftarrow{h_t} \qquad (4)$$

Prior Knowledge Integration Network. To control the process of learning effectively and enhance the sentence representation, we build a prior knowledge integration network. The structure is shown in Fig. 4.

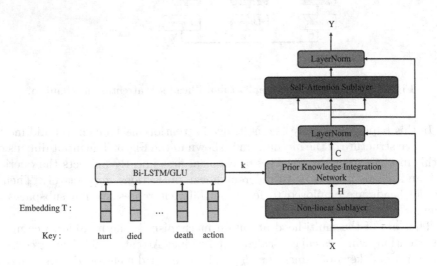

Fig. 4. A prior knowledge integration network is built in the model's first layer.

We build a keywords set based on ACE 2005 dataset, where keywords are high-frequency words as event triggers. The keywords sequence is $Key=key_1 key_2... key_m$, where key_i is one keyword in the collection and m is the total number of keywords. Looking up a pre-trained word embedding table to obtain the word embedding matrix T. Then we use the word embedding matrix T as the input of the neural network to learn the prior knowledge representation

k. In this model, we use Bi-LSTM and Gated Linear Unit (GLU) [4] to learn this representation respectively.

Bi-LSTM: Combine the last forward hidden state $\overrightarrow{h_m}$ and backword hidden state $\overleftarrow{h_1}$ to finally learn a prior knowledge representation k.

$$k = \overrightarrow{h_m} + \overleftarrow{h_1} \tag{5}$$

GLU: Given two filters $W \in \mathbb{R}^{l \times d \times d}$ and $V \in \mathbb{R}^{l \times d \times d}$, where d is the size of word embedding and l is the filter width. The output of GLU are computed as follows:

$$G = (T * W) \cdot \sigma(T * V) \tag{6}$$

where $G \in \mathbb{R}^{m \times d}$ is the output of GLU. We average the G to get the d-dimensional representation k.

Then we use k and the output of Non-linear Sublayer as the other input of prior knowledge network. The detailed process is: k is used as a query to compute the similarity with the output of Non-linear Sublayer H. After that we use the softmax function to get the normalized weights. Finally, combine the weights and H to get the result of filter C. To enhance the expression of sentence, we only add the prior knowledge integration network to the model's first layer.

$$e_t = v^T tanh(W_h h_t + W_k k) \tag{7}$$
$$\alpha_t = softmax(e_t) \tag{8}$$
$$c_t = \alpha_t * h_t \tag{9}$$

Position Encoding. Since self-attention mechanism cannot distinguish different positions. Therefore, it is important to encode the position of each word in the sequence. The position encoding method adopted in this model is using sine and cosine functions of different frequencies [19]:

$$PE(t, 2i) = sin(p/10000^{2i/d_{model}}) \tag{10}$$
$$PE(t, 2i + 1) = cos(p/10000^{2i/d_{model}}) \tag{11}$$

where t is the position and i is the dimension.

Loss Function. Given an input sequence $x = \{x_1, x_2, ..., x_n\}$, the log-likelihood of the corresponding correct label sequence $y = \{y_1, y_2, ..., y_n\}$ is

$$\log p(\boldsymbol{y}|\boldsymbol{x}; \boldsymbol{\theta}) = \sum_{t=1}^{n} \log p(y_t|\boldsymbol{x}; \boldsymbol{\theta}) \tag{12}$$

Our model obtain the conditional probability $p(y_t|\boldsymbol{x}, \theta)$ based on the vector $\boldsymbol{O_t}$ which is the output of the topmost attention sublayer:

$$p(y_t|\boldsymbol{x}, \boldsymbol{\theta}) = softmax(\boldsymbol{W_s O_t}) \tag{13}$$

where θ and $\boldsymbol{W_s}$ are parameters. Given all of our training set of size $|D|$, we can define the objective function as follows:

$$J(\boldsymbol{\theta}) = max \sum_{i=1}^{|D|} \log p(\boldsymbol{y}^{(i)}|\boldsymbol{x}^{(i)}, \boldsymbol{\theta}) \tag{14}$$

4 Experiment

4.1 Dataset

We conducted experiments on ACE 2005 dataset. ACE defines 8 event types and 33 subtypes. We randomly selected 30 articles from different genres as the development set, and subsequently conducted a blind test on a separate set of 40 ACE 2005 newswire documents. We used the remaining 529 articles as our training set.

4.2 Model Setup

Initialization. The weights of all sublayers are initialized as random orthogonal matrices. We initialize other parameters by sampling each element from a Gaussian distribution with mean 0 and variance $\frac{1}{\sqrt{d}}$. The embedding matrix is initialized randomly or using pre-trained word embedding.

Hyper-parameter Setting. We set the dimension of word embedding to 100 and the dimension of POS embedding to 100. The number of hidden layers is 4. The number of heads h is set to 8. The batch size is 32. To reduce overfitting, we apply dropout and label smoothing. Before residual connections we add dropout layers with a keep probability of 0.8. We also apply dropout layers before the attention softmax layer and the keep probability is set to 0.9. We set the smoothing value is 0.1 during training. The learning rate is initialized to 1.0. After training 40k steps, we halve the learning rate every 10K steps. We train all models for 160K steps.

4.3 Results on ACE 2005 Corpus

In this section, we conduct experiments on ACE 2005 corpus to demonstrate the effectiveness of the proposed approach. Firstly, we introduce systems implemented in this work.

S1 only relies on self-attention to complete the detection task.

S2 uses self-attention and prior knowledge integration to enhance the learning performance of the model. In our experiments, the size of keywords set is 150.

In Table 1, we give the comparisons of our model with previous approaches. Our approach without prior knowledge integration (S1) and with GLU-based

Table 1. Comparison with previous models on ACE 2005 dataset.

Models	Trigger(%)		
	P	R	F1
Li's joint model (2013)	73.7	62.3	67.5
Liu's PSL (2016)	75.3	64.4	69.4
Liu's FN-Based (2016)	77.6	65.2	70.7
Ngyuen's joint (2016)	66.0	73.0	69.3
Skin-CNN (2016)	N/A	N/A	71.3
Liu's ANN (2017)	78.0	66.3	71.7
S1	75.8	70.0	72.7
S2	**80.2**	71.5	**75.6**

prior knowledge integration (S2) achieves an F1 score of 72.7 and 75.6 respectively. Remarkably, S1 outperforms previous best performance by 1.0 F1 score. S2 has an improvement of 3.9 over the previous state-of-the-art and achieves the best performance. The experiment results prove that self-attention mechanism can help the model overcome the long-distance dependency problem and obtain the internal structure information of the sentence. At the same time, the prior knowledge integration adds control information during learning process to help reduce the interference of redundant information and further improve the performance of event detection.

4.4   Analysis

In this subsection, we analyze the main factors that influence our results on ACE 2005 dataset.

Table 2. Detailed results on ACE 2005 dataset.

#	Knowledge	Depth	F1
1	GLU	4	**75.6**
2	GLU	1	64.7
3	GLU	2	65.5
4	GLU	3	74.4
5	GLU	5	73.7
6	Bi-LSTM	4	67.5

Prior Knowledge Representation. The prior knowledge integration can help reduce redundant information and improve event detection. Row 1 and Row 6 in Table 2 compare two different prior knowledge representation learning methods. We can see that GLU can get better results.

Model Depth. The depth of models can affect the effect of model learning. Rows 1–5 of Table 2 show the effects of different number of layers. When the model with 1 layers only achieves 64.7 F1 score. Increasing depth consistently improves the performance on the development set, and our best model consists of 4 layers. When model with 5 layers, the performance drop of 1.9 F1.

5   Conclusion

In this work, we propose a novel approach to change event detection task to sequence labeling via self-attention mechanism. Besides, we also add the prior knowledge integration to control the process of learning, which improves performance according to the feature of event detection. In addition, we investigate different neural networks to learn the prior knowledge representation and compare experimental results. Our experimental results indicate that our models substantially improve event detection performance on ACE 2005 dataset.

References

1. Ahn, D.: The stages of event extraction. In: Proceedings of the Workshop on Annotating and Reasoning About Time and Events, ARTE 2006, pp. 1–8. Association for Computational Linguistics, Stroudsburg (2006). http://dl.acm.org/citation.cfm?id=1629235.1629236
2. Chen, Y., Xu, L., Liu, K., Zeng, D., Zhao, J.: Event extraction via dynamic multi-pooling convolutional neural networks. In: Proceedings of the 53rd Annual Meeting of the Association for Computational Linguistics and the 7th International Joint Conference on Natural Language Processing (Volume 1: Long Papers), pp. 167–176. Association for Computational Linguistics (2015). https://doi.org/10.3115/v1/P15-1017. http://www.aclweb.org/anthology/P15-1017
3. Cheng, J., Dong, L., Lapata, M.: Long short-term memory-networks for machine reading, January 2016
4. Dauphin, Y.N., Fan, A., Auli, M., Grangier, D.: Language modeling with gated convolutional networks. CoRR abs/1612.08083 (2016). http://arxiv.org/abs/1612.08083
5. Ji, H., Grishman, R.: Refining event extraction through cross-document inference. In: Proceedings of the Conference ACL 2008: HLT - 46th Annual Meeting of the Association for Computational Linguistics: Human Language Technologies, pp. 254–262 (2008)
6. Lei Ba, J., Kiros, J.R., Hinton, G.E.: Layer Normalization. ArXiv e-prints, July 2016
7. Li, Q., Ji, H., Huang, L.: Joint event extraction via structured prediction with global features. In: Proceedings of the 51st Annual Meeting of the Association for Computational Linguistics (Volume 1: Long Papers), pp. 73–82. Association for Computational Linguistics, Sofia, August 2013. http://www.aclweb.org/anthology/P13-1008
8. Liao, S., Grishman, R.: Using document level cross-event inference to improve event extraction. In: Proceedings of the 48th Annual Meeting of the Association for Computational Linguistics, ACL 2010, pp. 789–797. Association for Computational Linguistics, Stroudsburg (2010). http://dl.acm.org/citation.cfm?id=1858681.1858762

9. Nguyen, T.H., Cho, K., Grishman, R.: Joint event extraction via recurrent neural networks. In: Proceedings of the 2016 Conference of the North American Chapter of the Association for Computational Linguistics: Human Language Technologies, pp. 300–309. Association for Computational Linguistics (2016). https://doi.org/10.18653/v1/N16-1034. http://www.aclweb.org/anthology/N16-1034

10. Nguyen, T.H., Grishman, R.: Event detection and domain adaptation with convolutional neural networks. In: Proceedings of the 53rd Annual Meeting of the Association for Computational Linguistics and the 7th International Joint Conference on Natural Language Processing (Volume 2: Short Papers), pp. 365–371. Association for Computational Linguistics (2015). https://doi.org/10.3115/v1/P15-2060. http://www.aclweb.org/anthology/P15-2060

11. Parikh, A., Täckström, O., Das, D., Uszkoreit, J.: A decomposable attention model for natural language inference. In: Proceedings of the 2016 Conference on Empirical Methods in Natural Language Processing, pp. 2249–2255. Association for Computational Linguistics (2016). https://doi.org/10.18653/v1/D16-1244,. http://www.aclweb.org/anthology/D16-1244

12. Paulus, R., Xiong, C., Socher, R.: A deep reinforced model for abstractive summarization. CoRR abs/1705.04304 (2017). http://arxiv.org/abs/1705.04304

13. Pennington, J., Socher, R., Manning, C.D.: Glove: global vectors for word representation. In: Empirical Methods in Natural Language Processing (EMNLP), pp. 1532–1543 (2014). http://www.aclweb.org/anthology/D14-1162

14. Poon, H., Vanderwende, L.: Joint inference for knowledge extraction from biomedical literature. In: Human Language Technologies: the 2010 Annual Conference of the North American Chapter of the Association for Computational Linguistics, HLT 2010, pp. 813–821. Association for Computational Linguistics, Stroudsburg (2010). http://dl.acm.org/citation.cfm?id=1857999.1858122

15. Riedel, S., Chun, H.W., Takagi, T., Tsujii, J.: A Markov logic approach to biomolecular event extraction. In: Proceedings of the Workshop on Current Trends in Biomedical Natural Language Processing: Shared Task, BioNLP 2009, pp. 41–49. Association for Computational Linguistics, Stroudsburg (2009). http://dl.acm.org/citation.cfm?id=1572340.1572347

16. Riedel, S., McCallum, A.: Fast and robust joint models for biomedical event extraction. In: Proceedings of the Conference on Empirical Methods in Natural Language Processing, EMNLP 2011, pp. 1–12. Association for Computational Linguistics, Stroudsburg (2011). http://dl.acm.org/citation.cfm?id=2145432.2145434

17. Riedel, S., McCallum, A.: Robust biomedical event extraction with dual decomposition and minimal domain adaptation. In: Proceedings of the BioNLP Shared Task 2011 Workshop, BioNLP Shared Task 2011, pp. 46–50. Association for Computational Linguistics, Stroudsburg (2011). http://dl.acm.org/citation.cfm?id=2107691.2107698

18. Tan, Z., Wang, M., Xie, J., Chen, Y., Shi, X.: Deep semantic role labeling with self-attention. CoRR abs/1712.01586 (2017). http://arxiv.org/abs/1712.01586

19. Vaswani, A., et al.: Attention is all you need. CoRR abs/1706.03762 (2017). http://arxiv.org/abs/1706.03762

20. Zheng, S., Wang, F., Bao, H., Hao, Y., Zhou, P., Xu, B.: Joint extraction of entities and relations based on a novel tagging scheme. CoRR abs/1706.05075 (2017). http://arxiv.org/abs/1706.05075

Learning to Start for Sequence to Sequence Based Response Generation

Qingfu Zhu, Weinan Zhang, and Ting Liu[✉]

Research Center for Social Computing and Information Retrieval,
Harbin Institute of Technology, Harbin, China
{qfzhu,wnzhang,tliu}@ir.hit.edu.cn

Abstract. Response Generation which is a crucial component of a dialogue system can be modeled using the Sequence to Sequence (Seq2Seq) architecture. However, this kind of method suffers from vague responses of little meaningful content. One possible reason for generating vague responses is the different distribution of the first word between the generated responses and human responses. In fact, the Seq2Seq based method tends to generate high-frequency words in the beginning, which influences the following prediction resulting in vague responses. In this paper, we proposed a novel approach, namely learning to start (LTS), to learn how to generate the first word in the sequence to sequence architecture for response generation. Experimental results show that the proposed LTS model can enhance the performance of the start-of-the-art Seq2Seq model as well as other Seq2Seq models for response generation of short text conversation.

Keywords: Learning to start · Sequence to sequence
Response generation

1 Introduction

Recently, the sequence to sequence learning (Seq2Seq) architecture has gained great development as a general neural network method to model the potential relationship between two sequences. For the basic Seq2Seq model, each sequence is usually modeled by a recurrent neural network (RNN), and the two RNNs for the source sequence and target sequence are called encoder and decoder respectively. The encoder reads from the source sequence and does some summarization. The decoder is actually a language model that produces words according to previously predicted words conditioned with the encoder's output (usually called context vector). This indicates that when the decoder tries to predict a word, the context vector and the word predicted at last time are two necessary inputs.

However, the Seq2Seq method has an intrinsic problem in decoding: when producing the first word by the decoder, there is no previous predicted word to be referred to. Typically, previous works use a start symbol "</s>" to generate the

© Springer Nature Switzerland AG 2018
S. Zhang et al. (Eds.): CCIR 2018, LNCS 11168, pp. 274–285, 2018.
https://doi.org/10.1007/978-3-030-01012-6_22

first word [20]. While it is unreasonable to introduce an identical start symbol as the first word regardless of various source sequences. Concretely, there is no conditional probability of words can be learned. Meanwhile, the process of producing the first word and generating the rest words of a sequence is different so that they should be handled respectively.

To address the above issue, in this paper, we proposed a novel approach to learn to generate the first word in Seq2Seq learning process. In detail, we find two factors that may impact the decoding process: one is the representation of the source sequence which can be expressed using the encoder's states. The other is the representation of candidate words, whose information completely contained in the embedding matrix. We thus propose an independent neural network to model the two variables to optimize the generation of the first word in the decoding process and further improve the quality of the whole target sequence. The contribution of this paper is as follows:

- To the best of our knowledge, we are the first to propose an approach to learn to optimize the decoding of the first word in Seq2Seq architecture.
- The proposed LTS model can enhance the performance of the state-of-the-art Seq2Seq model as well as other Seq2Seq models for response generation of the short text conversation.
- Besides the short text conversation task, the proposed approach is a general framework which can also adapt to other Seq2Seq learning applications, such as machine translation, machine reading, text entailment, etc.

2   Method

In this section, we introduce the proposed Learning to Start Mechanism for response generation. As the Fig. 1 shows, the input message is first encoded into a context vector. The LTS then predicts the first word according to the context vector and the embeddings. Finally, it generates the rest of words starting with the first word, using the Seq2Seq model.

2.1   Encoding Process

The encoder takes a message as the input and summarizes it into a distribution representation, which is denoted as the context vector. Concretely, the encoder is modeled using a recurrent neural network (RNN) [11]. In this way, the encoding process can be described as:

$$h_i = f(h_{i-1}, m_i), \tag{1}$$

where h_i denotes the i-th hidden states of the encoder. m_i denotes the i-th word of the message. f is the recurrent unit of the encoder. Here, we use the Gated Recurrent Unit (GRU) [2] rather than the Long Short Term Memory (LSTM) [6] for all the RNNs in the proposed approach since the GRU has fewer parameters

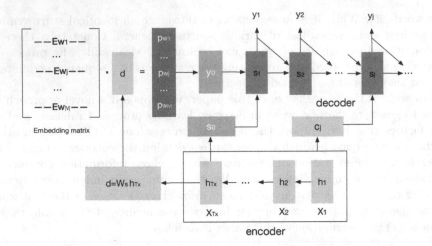

Fig. 1. An overview of our approach. Input sequence sent to the encoder is encoded into the context vector c_j, which can be either the last hidden state or a weighted sum according to whether attention mechanism is adopted. h_{T_x} which regarded as a summary of the message is used to initialize the s_0 and transferred by similarity matrix W_s into vector d in the embedding vector space at the same time. In this way, The probability distribution of the first word is then computed by the embedding matrix E and d.

than the LSTM. Given the hidden states, the encoder summaries them into the context vector through a function q.

$$c = q(h_1, h_2, \ldots, h_{T_x}), \tag{2}$$

where c is the context vector. T_x is the length of the message. The simplest form of q is directly taking the last hidden state:

$$c = h_{T_x}, \tag{3}$$

An alternative is to compute a weighted sum of all the hidden states, the weights can be calculated dynamically with the decoding process, which is known as the Attention Mechanism. It should be noted that the proposed approach is a universe mechanism for the Seq2Seq model, we thus combine it with different forms of q in our experiments to validate its effectiveness.

2.2   First Word Prediction

Given the context, the proposed approach predicts the first word from the vocabulary table. We cast it as a classification problem, where each word in the vocabulary table is regarded as a category, the LTS performs the prediction by computing a probability distribution over all categories and selecting the one with the highest probability. Concretely, we first represent a message using the last hidden state of the encoder:

$$rep_m = h_{T_x},\tag{4}$$

where $h_{T_x} \in \mathbb{R}^{d_h}$, d_h is the dimensionality of the hidden state. Besides, each word candidate w_i in the vocabulary is represented by its word embedding $e_{w_i} \in \mathbb{R}^{d_e}$, d_e is the dimensionality of the embeddings. In this way, we can compute a score for each word w_i using a similarity matrix $W_s \in \mathbb{R}^{d_h \times d_e}$.

$$score_{w_i} = rep_m \cdot W_s \cdot e_{w_i}.\tag{5}$$

The scores of all words are then normalized using the softmax function, resulting in the probability distribution of the first word:

$$p_{y_0} = \frac{exp(score_{w_i})}{\sum_{j=1}^{|V|} exp(s_{w_j})},\tag{6}$$

the word that has the highest probability is then selected as final prediction result:

$$y_0 = \arg\max p_{y_0}\tag{7}$$

2.3   Seq2Seq Prediction

Given the context vector and the first word, the reset words of a response are predicted using the Seq2Seq model. Similar to the encoder, the decoder is modeled using an RNN. In this way, the decoding process can be described as:

$$s_j = f(s_{j-1}, y_{j-1}, c_j),\tag{8}$$

where s_j is the j-th hidden state of the decoder. y_j is the j-th predicted word. c_j is the context vector for the j-th decoding time step. Noted that without the Attention Mechanism, c_j is a static context vector for different decoding steps, as Eq. 6 shows. For the Attention Mechanism, c_j is dynamically computed. At the j-th time step, for example,

$$c_j = \sum_{i=1}^{T_x} \alpha_{ij} h_i,\tag{9}$$

where α_{ij} is the weight of the i-th encoder hidden state. Concretely, α_{ij} is computed by:

$$\alpha_{ij} = \frac{exp(e_{ij})}{\sum_{k=1}^{T_x} exp(e_{kj})},\tag{10}$$

$$e_{ij} = a(s_{j-1}, h_i).\tag{11}$$

a is a forward neural network. Intuitively, the Attention Mechanism focuses on different parts of the message at different decoding time steps. This process can also be seen as an alignment.

Table 1. Some statistics of data sets. The training set is a one-to-one data set. The test set is a one-to-many data set, where a message corresponds to a varying number of responses. On average, each message has 42 responses in the test set.

Data	Data type	# of messages	# of responses
Training Data	one-to-one	1000,000	1000,000
Test Data	one-to-many	1000	42422

3   Experiments

In this section, we introduce the corpus in our experiments, baseline systems we compared with as well as parameter settings.

3.1   Data

We construct our training set by crawling from the Baidu Tieba[1], it is a Chinese forum where users can post and comment on others' posts. We extract roughly 1000,000 posts and their comments, regarding each post and its comment as the message and response, respectively.

It should be noted that our corpus is a one-to-one corpus, where each message corresponds to a single response. A similar public available corpus, NTCIR Corpus, used in [18] is one-to-many. Built by crawling from the Sina Weibo[2], each message in their corpus corresponds to a varying number of responses. We do not use the NTCIR Corpus as our training set due to the number of their messages is relatively small (roughly 190,000). But it well suits the requirement of BLEU evaluation because multiple responses have a more comprehensive coverage as references than a single response. We thus randomly select 1000 samples from NTCIR Corpus as our test set. Table 1 shows some statistics of these sets, and Table 2 shows some examples of our collected training set.

3.2   Baselines

The proposed approach is a universe mechanism to enhance the Seq2Seq based response generation models, we thus chose three state-of-the-art baselines then equip them with the LTS to validate the effectiveness of the proposed approach.

- **Vanilla Seq2Seq model.** The first baseline is the vanilla Seq2Seq model where both the encoder and the decoder are modeled using the RNN, and the context vector is the last hidden state of the encoder, as the Eq. 6 shows. We denote this baseline as Seq2Seq.
- **Attention based Seq2Seq model.** In this baseline, the context vector is computed dynamically at each decoding step [1], as the Eqs. 9 and 10 show. We denote this baseline as ATT.

---

[1] https://tieba.baidu.com/.
[2] https://weibo.com.

Table 2. Some examples of the training set. Each message corresponds to a single response.

message	response
今天天气好差呀	雨太大了
The weather is so bad today	The rain is too heavy
每天六点多出去	
打篮球锻炼身体	我在打网球
Get up at six everyday to play	I am playing tennis
basketball for exercise	
白色搭配什么颜色好	白色百搭呀
What color matches the white best?	White matches everything

- **Hybrid model.** Proposed by [18], this model combines the vanilla Seq2Seq model and the Attention based Seq2Seq by concatenating their context vectors as a new context vector. Intuitively, the Attention based model focuses on the local information of a message, while the vanilla Seq2Seq catches the global information. In this way, the hybrid model can take advantage of the two models by combining them together and training jointly. We denote this baseline as HYB.

3.3 Experiment Settings

All the baseline models and the proposed approach are implemented using the Keras[3]. We construct our vocabulary table using the most frequent 2,000 words of the training set. Any word not included in the table is replaced by a special token "UNK", which represents the unknown word. The word embeddings are pre-trained using the Word2Vec toolkit[4]. The word embedding dimensionality is set to 500. The number of units of both the encoder and the decoder is 1024. The batch size is 64. We use the Adam algorithm to optimize the parameters, the learning rate is set to 0.001.

4 Results

4.1 Evaluation Metrics

Until now, the evaluation of response generation is still an open question [4,15], we thus validate the effectiveness of the proposed approach using both human evaluation and automatic evaluation. For automatic evaluation, we first test the proposed approach by some statistics of the first word prediction. Besides, we also evaluate a complete response using BLEU [13] to see whether the proposed approach can improve the performance of the Seq2Seq model.

---

[3] https://keras.io/.
[4] https://code.google.com/archive/p/word2vec/.

Table 3. Some examples of human evaluation metrics.

post	response	score
咳嗽不止，有啥良策吗 Can not stop cough, any good idea?	吃点退烧药 Eat some antipyretics	0
永远不要对一个外行聊你的专业 Never talk about your major with strangers	我很专业的 I'm very professional	0
自制中号三明治 Medium sandwich made by myself.	我也想吃 I want some, too	+1
哪本新华字典是你用过的 Which XinHua dictionary is the one you used	两本都是 Both of them are	+1
完事了...准备去吃饭，明天回家 All fixed... ready for dinner, go back home tomorrow	还没吃饭啊？ Haven't eaten yet?	+2

First Word Prediction Evaluation Metrics. We evaluate the generation of the first word from two aspects: accuracy and diversity. For each test sample, we define its "R-set" as the set of the first word in all references. During the test process, once the first predicted word is in the R-set, then the test sample is marked as a hit. The accuracy is defined as the number of hits normalized by the total number of test samples.

However, the basic accuracy can be badly influenced by high-frequency words due to they widely exist in references. In this way, a response can easily hit its R-set as long as its first word is a high-frequency word. We thus propose the "accuracy without top k frequency word" (denoted as $accw$-k) based on the definition of the accuracy. In $accw$-k, the top k high-frequency words (counted on the training set) are removed from the R-set of each sample. Specially, $accw$-0 is exactly the basic accuracy that does not neglect any high-frequency words. For example, suppose the first predicted word of a sample is "I", its complete R-set is { "he", "I", "eat"}. The top two high-frequency words are "the" and "I", respectively. The sample will be labeled as a hit in the $accw$-0 and $accw$-1. But for the following $accw$-k where k is larger than 2, "the" and "I" are removed from the R-set, the sample will not be regarded as hit anymore.

We also propose another metric to evaluate the diversity: div-k metric. It is defined as the ratio of test samples whose first word is not in top k high-frequency words.

Human Evaluation. We also validate the proposed approach via human evaluation. The evaluation metrics are made referring to [18]. Three annotators[5] are employed to score responses in a range of 0 to 2. The detailed metrics are as follows:

- 0: This indicates a bad response. If a response has grammar or fluency mistakes, is not logically consistent or semantically relevant to the message, it should be labeled as 0.

---

[5] All annotators are well-educated students and have a Bachelor or higher degree.

Table 4. Results of BLEU evaluation.

Model	BLEU-1	BLEU-2	BLEU-3	BLEU-4
Seq2Seq	42.870	3.226	0.382	0.000
Seq2Seq+LTS	**46.905**	**4.138**	**0.433**	**0.100**
ATT	**48.030**	3.884	0.450	**0.100**
ATT+LTS	45.634	**4.496**	**0.799**	**0.100**
HYP	**47.923**	4.403	0.516	0.000
HYP+LTS	47.488	**4.404**	**0.526**	**0.225**

- +1: This means the response may not perfect, but is suitable for some particular scenario.
- +2: This indicates a quite appropriate response without grammar or fluency errors and is independent of scenario.

The annotations of different responses to a message are independent. For example, they can all be labeled as 0 if none of them is a suitable response. Table 3 shows some examples of the annotation metrics. The first example conflicts with the logic consistency principle, the message says he gets a cough, but the response advises antipyretics, which is not logically relevant. The second example is not semantically relevant to its message. The response in the third example can be seen as a suitable one but it is too generic, leading to a score of one. The fourth response strongly depends on a particular scenario that the author of the message must have exactly two dictionaries. The last example shows a suitable response without these problems.

4.2 Results and Analysis

The results of BLEU evaluation are shown in Table 4. The proposed approach significantly outperforms baselines from BLEU-2 to BLEU-4. This indicates the LTS is helpful for the Seq2Seq based models to generate more longer n-grams that appear in references.

The evaluation of the first word accuracy is shown in Fig. 2. The baselines models in three groups have a higher accuracy at $accw$-0. But their accuracies decline rapidly when high-frequency words are ignored since $accw$-1. LTS models have more stable curves which are less insensitive to the high-frequency words. Figure 3 shows evaluation results of the first word diversity. Similar to the accuracy metric, the diversity of baseline models declines more rapidly than the proposed approach with the increase of k. Especially for the Seq2Seq baseline, there is a significant reduction when the top one frequency word is ignored. We believe this is due to the relatively simple structure of its context vector. However, by combining with the LTS, the decay speed slows down significantly. This indicates that the LTS is effective in improving the diversity of the first word prediction.

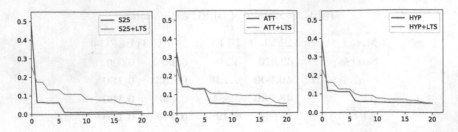

Fig. 2. Accuracy of first word prediction. The y-axis and x-axis of each plot correspond to accw-k (the accuracy of neglecting top k high-frequency words) and position k respectively.

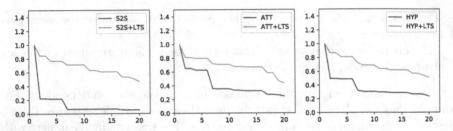

Fig. 3. Diversity of first word prediction. The y-axis and x-axis of each plot correspond to div-k (the ratio of the first predicted word is not in top k high-frequency words) and position k respectively.

Table 5. Human evaluation results about mean scores and proportions of three rating levels.

Models	Mean	0	+1	+2	kappa
Seq2Seq	1.20	72.66%	14.66%	12.66%	0.309
Seq2Seq+LTS	**1.24**	71.33%	16.00%	12.66%	0.332
ATT	1.16	74.66%	12.00%	13.33%	0.398
ATT+LTS	**1.26**	68.66%	20.66%	10.66%	0.269
HYP	1.11	73.33%	16.66%	10.0%	0.340
HYP+LTS	**1.20**	70.00%	20.00%	10.00%	0.333

Human evaluation results are shown in the Table 5. The proposed approach outperforms all three baselines in the mean score. Agreements among different annotators are calculated by Fleiss' Kappa [3]. All the kappa values are in a range of 0.2 to 0.4, which can be seen as "Fair agreement". Table 6 shows some examples of responses generated by different models.

5   Related Work

Early response generation systems are often built using rule-based [22] or learning-based methods [10,16,24]. These systems require extensive human

Table 6. Some response examples of the baseline models and our proposed model.

No.	Message	Model	Response	Translated Response
1	想改个名字，大家出个主意呗 Want to change a name, has anyone got any idea	Seq2Seq	不想啥呀	Don't want anything
		Seq2Seq+LTS	我不想改	I don't want to change it
		ATT	不想改了	Don't want to change any more
		ATT+LTS	那就改呗	Then change it
		HYP	我改不了	I can not change it
		HYP+LTS	别乱改呀	Don't change it casually
2	再不疯狂，我们就老了 We grow old by deserting unrestrained	Seq2Seq	昨天就疯了	Have been crazy since yesterday
		Seq2Seq+LTS	本来就老了	Already old.
		ATT	你也老了	You are old, too
		ATT+LTS	我们都老了	We all grow older
		HYP	你也疯狂啊	You are unrestrained, too
		HYP+LTS	人老心不老	Old in age, but young in mind
3	这太空漫步太猛了... The space is so awesome...	Seq2Seq	这也能凉快	It must be cool
		Seq2Seq+LTS	雨中漫步吧	Let's walking in the rain
		ATT	漫步也是吧	Walking is in the same way
		ATT+LTS	是挺猛的	It's pretty awesome
		HYP	雨太大了	The rain is too heavily
		HYP+LTS	是挺猛的	It's pretty awesome

efforts to make the rules, making it difficult to scale up. Recently, as the blooming of the neural network, the sequence to sequence model has gained promising performance in text generation tasks, such as machine translation [1,2,20], document summarization [14] as well as response generation [19,21]. Based on the standard Seq2Seq model, [18] introduce the Attention Mechanism and further propose a hybrid model for response generation. However, the Seq2Seq based response generation model suffers from the generic response problem that most responses are vague and contain little substantial contents, such as "I do not know", "Me, too".

Research to address this issue can be roughly divided into two categories. The first type is to optimize the training process of the model. [7] introduce the Maximum Mutual Information (MMI) loss as the objective function. [8] cast the response generation as a reinforcement learning problem. The reward is provided by some reward functions, which can be designed to punish the generic responses. However, the evaluation of responses is still an open question, it is hard to design a comprehensive reward function. To this end, [9] introduce a discriminator to estimate the reward borrowing the idea from the adversarial training [5]. Another way to address the generic response problem is to introduce external knowledge. [12] divide the process of response generation into two steps. First, they predict a keyword using the Point Mutual Information (PMI). They then generate the rest of the response starting with the keyword. Similarly, [23] first predict multiple topic words, then condition them to the decoder using the attention mechanism to generate topic-aware responses. [17] also proposed a multi-keywords based model. Different from [23], keywords in their model are extracted from the message, and condition to the encoding process using an extra keyword encoder.

6 Conclusions

In this paper, we proposed a novel approach to optimize the generation of the first word in the decoding process of Seq2Seq models for response generation of short text conversation. Experimental results show that the proposed LTS model can enhance the performance of the start-of-the-art model as well as other Seq2Seq models.

References

1. Bahdanau, D., Cho, K., Bengio, Y.: Neural machine translation by jointly learning to align and translate. arXiv preprint arXiv:1409.0473 (2014)
2. Cho, K., Gulcehre, B.v.M.C., Bahdanau, D., Schwenk, F.B.H., Bengio, Y.: Learning phrase representations using rnn encoder-decoder for statistical machine translation. In: Proceedings of the 2016 Conference on Empirical Methods in Natural Language Processing (2014)
3. Fleiss, J.L.: Measuring nominal scale agreement among many raters. Psychol. Bull. **76**(5), 378 (1971)
4. Galley, M., et al.: deltaBLEU: a discriminative metric for generation tasks with intrinsically diverse targets. In: Proceedings of the 53rd Annual Meeting of the Association for Computational Linguistics and the 7th International Joint Conference on Natural Language Processing, vol. 2, pp. 445–450 (2015)
5. Goodfellow, I., et al.: Generative adversarial nets. In: Advances in Neural Information Processing Systems, pp. 2672–2680 (2014)
6. Hochreiter, S., Schmidhuber, J.: Long short-term memory. Neural Comput. **9**(8), 1735–1780 (1997)
7. Li, J., Galley, M., Brockett, C., Gao, J., Dolan, B.: A diversity-promoting objective function for neural conversation models. In: Proceedings of the 2016 Conference of the North American Chapter of the Association for Computational Linguistics: Human Language Technologies, pp. 110–119 (2016)
8. Li, J., Monroe, W., Ritter, A., Jurafsky, D., Galley, M., Gao, J.: Deep reinforcement learning for dialogue generation. In: Proceedings of the 2016 Conference on Empirical Methods in Natural Language Processing, pp. 1192–1202 (2016)
9. Li, J., Monroe, W., Shi, T., Jean, S., Ritter, A., Jurafsky, D.: Adversarial learning for neural dialogue generation. arXiv preprint arXiv:1701.06547 (2017)
10. Litman, D., Singh, S., Kearns, M., Walker, M.: NJFun: a reinforcement learning spoken dialogue system. In: Proceedings of the ANLP-NAACL 2000 Workshop on Conversational Systems, pp. 17–20. Association for Computational Linguistics (2000)
11. Mikolov, T., Karafiat, M., Burget, L., Cernocky, J., Khudanpur, S.: Recurrent neural network based language model. In: Eleventh Annual Conference of the International Speech Communication Association (2010)
12. Mou, L., Song, Y., Yan, R., Li, G., Zhang, L., Jin, Z.: Sequence to backward and forward sequences: a content-introducing approach to generative short-text conversation. In: Proceedings of COLING 2016, the 26th International Conference on Computational Linguistics: Technical Papers, pp. 3349–3358 (2016)
13. Papineni, K., Roukos, S., Ward, T., Zhu, W.J.: Bleu: a method for automatic evaluation of machine translation. In: Proceedings of the 40th annual meeting on association for computational linguistics, pp. 311–318. Association for Computational Linguistics (2002)

14. Rush, A.M., Chopra, S., Weston, J.: A neural attention model for abstractive sentence summarization. arXiv preprint arXiv:1509.00685 (2015)
15. Schatzmann, J., Georgila, K., Young, S.: Quantitative evaluation of user simulation techniques for spoken dialogue systems. In: 6th SIGdial Workshop on DISCOURSE and DIALOGUE (2005)
16. Schatzmann, J., Weilhammer, K., Stuttle, M., Young, S.: A survey of statistical user simulation techniques for reinforcement-learning of dialogue management strategies. Knowl. Eng. Rev. **21**(2), 97–126 (2006)
17. Serban, I.V., et al.: Multiresolution recurrent neural networks: an application to dialogue response generation. In: AAAI, pp. 3288–3294 (2017)
18. Shang, L., Lu, Z., Li, H.: Neural responding machine for short-text conversation. In: Proceedings of the 53rd Annual Meeting of the Association for Computational Linguistics and the 7th International Joint Conference on Natural Language Processing, vol. 1, pp. 1577–1586 (2015)
19. Sordoni, A., et al.: A neural network approach to context-sensitive generation of conversational responses. In: Proceedings of the 2015 Conference of the North American Chapter of the Association for Computational Linguistics: Human Language Technologies, pp. 196–205 (2015)
20. Sutskever, I., Vinyals, O., Le, Q.V.: Sequence to sequence learning with neural networks. In: Advances in Neural Information Processing Systems, pp. 3104–3112 (2014)
21. Vinyals, O., Le, Q.: A neural conversational model. arXiv preprint arXiv:1506.05869 (2015)
22. Weizenbaum, J.: Eliza–a computer program for the study of natural language communication between man and machine. Commun. ACM **9**(1), 36–45 (1966)
23. Xing, C., et al.: Topic aware neural response generation. AAAI **17**, 3351–3357 (2017)
24. Young, S., Gašić, M., Thomson, B., Williams, J.D.: Pomdp-based statistical spoken dialog systems: a review. Proc. IEEE **101**(5), 1160–1179 (2013)

Author Index

Printed in the United States
By Bookmasters